Electrical and Instrumentation Safety for Chemical Processes

Electrical and Instrumentation Safety for Chemical Processes

Richard J. Buschart

VNR VAN NOSTRAND REINHOLD
New York

Softcover reprint of the hardcover 1st edition 1991

Library of Congress Catalog Card Number 91-572
ISBN-13: 978-1-4684-6622-5

Published by Van Nostrand Reinhold
115 Fifth Avenue
New York, New York 10003

Chapman and Hall
2-6 Boundary Row
London, SE1 8HN, England

Thomas Nelson Australia
102 Dodds Street
South Melbourne 3205
Victoria, Australia

Nelson Canada
1120 Birchmount Road
Scarborough, Ontario M1K 5G4, Canada

16 15 14 13 12 11 10 9 8 7 6 5 4 3 2 1

Library of Congress Cataloging-in-Publication Data

Buschart, Richard J.
Electrical and instrumentation safety for chemical processes / Richard J. Buschart
p. cm.
Includes bibliographical references and index.
ISBN-13: 978-1-4684-6622-5 e-ISBN-13: 978-1-4684-6620-1
DOI: 10.1007/978-1-4684-6620-1
1. Chemical plants—Electric equipment—Safety measures.
I. Title.
TP157.B87 1991
660'.2804—dc20

This text is dedicated to those involved in the design, operation, maintenance, and management of chemical and similar processes, with the goal that it will help them provide the safest facilities possible, and to those who live near chemical process facilities.

It is also dedicated to those who do not know what they do not know. In some chemical process units, there are no experienced electrical engineers, and someone else will have to make electrical safety decisions. This book will help that person to know what he or she does not know.

Contents

Preface

This text is about electrical and instrumentation safety for chemical processes. It covers a wide area of electrical and electronic phenomena and how they have and can significantly affect the safety of chemical processes. The importance of the subject is well known to anyone involved in the operation of chemical processes.

Lightning strikes can explode storage tanks, shut down electrical power systems, and shut down or damage computer and instrument systems. Static electricity can ignite flammable materials and damage sensitive electronic process control equipment. Electrical power system failures or interruptions can produce unsafe process conditions. Chemical processes use flammable and combustible vapors, gases, or dusts that can be exploded by electrical equipment and wiring. Even low-energy equipment like flashlights can ignite a flammable vapor. Interlock and equipment protection systems can cause safety problems.

How important is electrical and process control safety? A survey on "How Safe is Your Plant?", in the April 1988 issue of *Chemical Engineering* magazine, provided some answers. Among the results of this survey of chemical processes, it was found that over 800 respondents believed instrumentation and controls, shutdown systems, equipment interlocks, and other protection systems to be the least safe aspect of chemical industries. The survey also indicated that complying with OSHA and other regulations, process control software security, inspections, audits, and safety training are important safety issues.

Electrical and Instrumentation Safety for Chemical Processes covers the areas mentioned in the *Chemical Engineering* magazine survey, as well as many other important issues, including: hazardous areas; combustible dust safety; computer power and grounding; power system reliability for chemical processes; and process control safety.

Electrical and Instrumentation Safety for Chemical Processes

1

Introduction

The mission of this text is to present the principles and practice of electrical and process control safety as applied to chemical processes. The text accomplishes the following:

1. It presents the wide spectrum of electrical and process control areas essential to safety.
2. It presents the topics in an integrated approach that ties together key issues.
3. It takes unique and creative approaches to controversial issues.
4. It provides a safety course with practical, usable criteria. The text is prepared from actual experiences and work in chemical processes.
5. It presents a discussion, based on personal experience with work on codes and standards committees and with their application to new facilities and modifications of existing plants, of national and international regulations and standards; it describes what they are about and how to comply with them.
6. It provides recommendations concerning documentation, training, safety audits, inspections, and accident investigations.
7. It discusses the application of new techniques such as combustible gas detectors, fault-tolerant systems, and European hazardous area practices.
8. This test is a guide to good safety practice; however, applying the principles and practices in day-to-day situations requires good judgment and careful consideration of the factors involved. Alternatives and modifications should always be considered where justified and appropriate.
9. The National Fire Protection Association (NFPA), the Instrument Society of America (ISA), and other organizations' codes and standards are referred to throughout this text. It is essential to refer to the wording in the current code or standard in any situation.
10. The principles and practices described in this text are believed to

be safe and good practice and are intended to be used with proper engineering practice and application. Since there is a variety of conditions that can exist in any given situation that are beyond our control, the author, the author's employer, and the publisher assume no liability in connection with the application of the principles and practices described herein.

GENERAL SAFETY CRITERIA: PROTECTION OF PEOPLE, PROPERTY, AND THE COMMUNITY

The following is an excerpt from an article in the December 1988/January 1989 issue of *Chemical Manufacturers Association* (*CMA*) *News* titled "A Blueprint for Regaining Public Trust," which indicates that "in the following excerpts from a recent speech before the American Institute of Chemical Engineers, Harold J. Corbett, Senior Vice President of Monsanto Company, describes what industry must do to regain public trust."

The excerpt from the speech states that

> process risks will be reduced to the point where serious accidents do not occur and people living near the plant site will fully understand how the plant operates. Moreover they will have full confidence in the people operating the plant because they know that these people make safety the top priority.

The results of a survey on "How Safe is Your Plant?" in the October 10, 1988, issue of *Chemical Engineering* indicate that 90% of the respondents feel safe on the job, despite the fact that 44% of their plants had a major accident within the last 5 years and 70% think a similar accident may happen again. Safety is a major concern in all industries but especially the chemical process industry because of the hazardous nature of the materials that are used. It is absolutely essential to avoid catastrophic incidents like Flixborough. Flixborough refers to an explosion that occurred at a caprolactam facility in the United Kingdom in 1974. The explosion occurred when 50 tons of cyclohexane was released into the air, killing 28 employees, injuring many, and causing $100 million in property damage. Such incidents like this and the 1989 explosion at a Phillips Petroleum plastic plant in Pasadena, Texas, could result in additional regulatory legislation on the state and federal level. In many states, hazop analyses are already a part of state regulations for certain high hazard materials. Recent oil spills have resulted in significant federal and state legislation to regulate storage tanks and pipelines.

It is important to recognize that environment and safety issues share a common ground. Explosions, fires, and malfunctions of process equipment can cause the release of materials outside the plant boundary. Furthermore, public perception and opinion is critical in today's regulatory environment.

Fires, explosions, injuries, and fatalities within the plant that are not in the Flixborough category are equally damaging. Each incident, no matter how large or small, is another bad mark against the plant, company, and industry. The individual and aggregate impact on the individual process unit, plant, and company can be significant. Safety is good business. This is as true today as it has ever been, especially in the competitive environment. Companies cannot afford accidents. Fires and explosions can render a process unit or plant inoperative. The replacement of facilities at today's inflated costs may be too high to justify replacement, and the unit may be shut down. Lost production results in lost sales and unhappy customers who may buy from competitors.

As an example, a product testing lab had a fluorescent lighting fixture fire (not an uncommon event) that shut down the lab, and the product had to be shipped outside the plant to be tested. As another example, a static spark ignited a cotton-like plastic that burned the building. The unit was shut down permanently because rebuilding was not justified. A fire in a process control room or computer room can be particularly devastating because it can result in the loss of process control or interlock design that may not be recoverable.

Personnel injuries or fatalities are particularly devastating, especially if they involve fellow employees and friends. In addition to personal tragedy, which is bad enough, Occupation and Safety Health Administration (OSHA) citations and lawsuits can follow. The average cost of a disabling injury in 1987 was approximately $23,000, which represents a 22% increase over 1986.

Personnel injuries or fatalities can occur in electrical systems without explosions or fires. Thirty volts AC (or in some situations even lower voltage) can electrocute a person. For example, an improperly grounded limit switch or instrument transmitter could be the cause. As another example, an electric arc is the hottest temperature on earth next to a nuclear reaction. A spark between the wiring terminals in a control panel could damage the eyesight of an electrician looking closely at the source to troubleshoot a circuit but not wearing safety glasses. Arcs in power equipment, which can release significant thermal energy, have burned maintenance electricians severely.

If the costs and consequences of accidents in chemical processes are understood, why do they occur? The following are possible reasons.

Production and Costs

The safety survey in *Chemical Engineering* magazine (October 1988) indicates that although safety is often said to be paramount, in reality production often reigns in many Chemical Process Industries plants. One respondent indicated that safety is talked about, but production and costs are always the important factors.

It Can't Happen Here

The tendency to think that an accident cannot happen in your plant is not peculiar to chemical process safety. It might be associated with a failure to appreciate what can happen and inexperience with the fact that it *has* happened here. Dusts are an example of something that is familiar and appears to exist in a rather harmless condition in day-to-day operations. In certain conditions, however, dust can produce violent explosions and fires, as grain-processing people can testify.

Poor Design and Installation

Good engineering is an essential element of chemical process safety, or any type of safety. Well-engineered plants produce the quality and quantity of products needed in a safe manner. They can be operated or maintained without undue effort or safety risks. They start up and operate well.

Operator Error

In many instances, operator error is cited as the reason for an accident. Operator error may be the result of inadequate training or supervision, but the control system design may be a contributing factor. The design needs to be friendly, easy to operate, not overly complicated, and without too many alarms or interlocks (overalarming and interlocking are common maladies of control system design). The process information needs to be presented in an easy-to-understand and usable manner. Controls need to be designed to account for equipment and control system failures and operator error.

Inadequate Maintenance

Inadequate maintenance has also been cited as a contributing factor to accidents. Some maintenance problems may be attributable to poor design and installation. The equipment may be inaccessible or may not have adequate working space. [These deficiencies are violations of the OSHA regulations and the *National Electrical Code* (*NEC*).] (*National Electrical Code®* and *NEC®* are registered trademarks of the National Fire Protection Association, Inc., Quincy, MA 02269.) The equipment may be difficult to calibrate or test; the proper test or calibration equipment may not have been provided; or the vendor data may be nonexistent, in error, or too complicated.

There are undoubtedly other reasons or contributing factors as to why accidents occur, but regardless of the factors, they could have been controlled.

SAFETY PHILOSOPHY AND PRINCIPLES

The following excerpt is from the Bureau of Naval Personnel Rate Training Manual, *Basic Electricity:*

> In the performance of his normal duties, the technician is exposed to many potentially dangerous conditions and situations. No training manual, no set of rules or regulations, no listing of hazards can make working conditions completely safe. However, it is possible for the technician to complete a full career without serious accident or injury. Attainment of this goal requires that he be aware of the main sources of danger, and that he remain constantly alert to those dangers. He must take the proper precautions and practice the basic rules of safety. He must be safety conscious at all times, and this safety consciousness must become second nature to him.

The same statement is true for those who work in chemical processes. There are people who have worked more than 40 years in chemical facilities without being involved in an accident or sustaining any injury at work.

Explosions, fires, spills, releases, injuries, and fatalities can all be prevented. A statement from the cover of the Chemical Safety Data Sheets published by the Manufacturing Chemists Association (now the Chemical Manufactures Association) indicates that chemicals in any form can be handled, stored, or processed if the physical, chemical, and hazardous properties are fully understood and the necessary precautions, including the use of protective equipment and proper safeguards are followed.

Webster's definition of an accident includes the phrases "an event occurring by chance or from unknown causes" and "any unfortunate event resulting from carelessness, unawareness, ignorance, or unavoidable causes." The second statement describes the type of situation with which we are concerned, except we should question "unavoidable causes." Nearly all accidents are avoidable. Some accident reports seem to indicate that the cause of the accident was mysterious forces, unforeseen conditions, or an act of God.

My experience with accident investigation indicates that this is not true. In most cases, established codes and practices, standards, and sound engineering principles have been violated. In a court case involving lightning damage to a structure, an attorney argued that lightning has for some time been considered "an unforeseeable intervening force," but it is a "reasonably foreseeable danger" for which there are inexpensive simple protection systems. (Lightning protection systems were proposed by Ben Franklin.)

The same is true for electrical and process control safety in chemical processes. The engineering technology to protect against hazards is known; however, it is not always simple or inexpensive.

The steps in achieving safety are as follows:

1. Know the nature and characteristics of the hazard involved.
2. Convert this knowledge to usable, understandable, practical information. This information includes reference documentation, standards and practices, technical studies and papers, and training and inspection material.
3. Provide training—especially at the operating and maintenance level.
4. Conduct inspections, audits, and investigations to provide feedback to the organization.
5. Update and upgrade the process as technology, people, and the environment change.

Electrical and control system safety requires interaction and coordination with many other systems, disciplines, and groups of people. Structural, mechanical, and process design, plant operations, plant maintenance, construction, and purchasing are all part of the process, as are staff safety and research. Safety involves a systems approach. Failure to recognize this requirement can only lead to less than desirable results.

ELECTRICAL AND CONTROL SYSTEM SAFETY INCIDENTS

The following is a list of some safety incidents that have occurred, some a number of times, and can be expected to happen again:

1. Lightning that struck caused an explosion in a flammable liquid storage tank.
2. Static discharge exploded vapor in a duct system.

3. A building exploded as a result of interlock bypass and runaway reaction.
4. Programmable Logic Controller (PLC) and Unit Operations Controller locked up during an electrical storm.
5. Thermowell failure caused toxic and flammable gas to enter a control room.

ACRONYMS AND ABBREVIATIONS

The following acronyms and abbreviations are used throughout this text.

AIChE American Institute of Chemical Engineers
AIT Autogenous ignition temperature
ANSI American National Standards Institute
API American Petroleum Institute
ASME American Society of Mechanical Engineers
ASTM American Society for Testing Materials
AWG American wire gauge
CEC *Canadian Electrical Code*
CENELEC European Electrotechnical Committee for Standardization
CRT Cathode ray tube
CMA Chemical Manufacturers Association
CSA Canadian Standards Association
CT Current transformer
DCS Distributed Control Systems
DIP Dust-ignition-proof
d/p Differential pressure
EMI Electromagnetic interference
EP Explosionproof (enclosure)
EPA Environmental Protection Agency
FM Factory Mutual Research Corporation
FP Flash point
GFCI Ground Fault Circuit Interrupter
HP Horsepower
HS Hermetically sealed (contact)
I Amperes
IEC International Electrotechnical Commission
IEEE Institute of Electrical and Electronics Engineers
IES Illuminating Engineering Society
I/O Input/Output
IS Instrinsic safety
ISA Instrument Society of America

KVA	Kilovolt ampere
KW	Kilowatts
LEL	Lower explosive limit (lower flammable limit)
LNG	Liquified natural gas
LPG	Liquified petroleum gas
MC	Metal clad (cable)
MCC	Motor control center
MESG	Maximum experimental safe gap
MG	Motor-generator set
MI	Mineral insulated (cable)
MOV	Metal oxide varistors
MV	Medium voltage
NAS	National Academy of Science
NEC	*National Electrical Code*
NEMA	National Electrical Manufacturers Association
NFPA	National Fire Protection Association
NI	Nonincendive (circuit or component)
NRC	National Research Council
OSHA	Occupational Safety and Health Administration
PCB	Polychlorinated biphenyls
P and ID	Process and instrumentation diagram
PES	Programmable Electronics Systems
PLC	Programmable Logic Controller
PLTC	Power limited tray cable
PP	Purging and pressurization (enclosure)
PWM	Pulse-width modulation
R	Resistance
RFI	Radio frequency interference
RTD	Resistance Temperature Detectors
SCR	Silicon Controlled Recifier
SIT	Spontaneous ignition temperature
SPS	Standby power supply
TC	Tray cable
TCD	Technical Committee Documentation
TCR	Technical Committee Report
TEFC	Totally enclosed fan-cooled (motor)
UEL	Upper explosive limit (upper flammable limit)
UL	Underwriters Laboratories, Inc.
UPS	Uninterruptible power supply
V	Voltage
Z	Impedance
WPII	weather protected II

2

Characterization of Process Units for Electrical and Process Control Safety

It is useful to characterize a particular process unit according to the distinctive and unique features that influence its electrical and control system safety. This characterization can provide the basis for approaching many safety questions, such as What type of hazardous classification may be required? What type of interlocks may be needed? What type of protective systems may be required? How sensitive is the process to power disturbances and failures? What type of special precautions and preventive measures should be considered? Dow's *Fire and Explosion Index Hazard Classification Guide* (fifth edition), published by Chemical Engineering Progress and the American Institute of Chemical Engineers (AIChE) (1981), provides a useful reference for explosion and fire safety. It is directed at the risk of facility and business interruption losses from fire and explosion. It does not, however, cover the total safety picture; that is, it does not address individual personnel safety or environmental concerns.

PROCESS CONDITIONS THAT INFLUENCE SAFETY

Process materials are an important factor in determining the relative safety risks of one process unit versus another. Consideration should be given to the process materials in all steps and phases of the operation under normal and abnormal operation. This includes raw materials, feed stocks, intermediates, final products, waste streams, and catalysts. Chlorine, hydrogen cyanide, and other high hazard materials require high levels of safe guards in their electrical and control system facilities.

A fire and explosion index can be calculated using the Dow guide for a particular process. It includes the following:

Process temperature above the flash points
Process temperature above the boiling points
Process temperature above the autoignition temperature
Operation in or near the flammable range

This requires knowledge of the fire hazard properties of the particular process streams. Please note that some streams do not consist of pure components and require calculation or measurement of these properties. For instance, the flash point and flammability limits of aqueous solutions of flammable liquids depend on the compositions of water and the flammable liquid and the temperature of the solution.

The flammable properties for gases, vapors, and liquids are

Flash point
Autoignition temperature
Flammability limits
Flammability range
Boiling point
Specific gravity, as a gas or vapor
NEC Explosion Group A, B, C, or D

These values are available in various NFPA publications as well as in *Dangerous Properties of Industrial Materials* by N. Irving Sax.

Toxicity, including safe health exposure levels and environmental restrictions, are also factors that influence the safety requirements for any facility. OSHA, Environmental Protection Agency (EPA), and state and federal regulations should be defined for any facility.

The electrical conductivity and dielectric constant need to be known for liquids in order to determine the potential for generation of static electricity. Corrosive materials have a greater potential for equipment failure. Therefore, electrical and instrument enclosures must be constructed of materials that will not be corroded by process fluids. Electrical cable tray and conduit systems must be compatible with the environment. This is very important.

Electrical pipe heat-tracing systems have failed and have caused fires. Therefore, if there are materials like caustic or formaline that may require pipe tracing, these requirements need to be identified. The Chemical Safety Data Sheets provide an excellent resource for many materials and a check list for new materials. The items in the data sheets include

Flammability limits
Flash point: closed cup: open cup
Autoignition temperature
Boiling point

Coefficient of expansion
Corrosivity
Critical pressure
Critical temperature
Density
Heat of combustion
Heat of fusion
Hygroscropicity
Melting point
Odor
Reactivity
Solubility
Specific gravity
Specific heat
Vapor density
Vapor pressure
Viscosity
Threshold limit in air
Health hazards
Fire and explosion hazards
Building design
Equipment design
Ventilation
Electrical equipment
Static safety
Employee safety
Personal protective equipment
Fire fighting
Handling and storage
Tank and equipment cleaning and repairs
Waste disposal

PROCESS OPERATIONS

The process operations, including the basic chemistry, process controllability, how well it responds to disturbances, and the type of response all influence the safety requirements for the electrical and control system. The type of reaction (exothermic or endothermic), possibility of a runaway reaction, sensitivity to equipment failure, interlock requirements, process conditions, process temperature above the flash point, operation within the flammability range, batch versus continuous operations, material handling and manual loading or unloading operations all influence the process safety requirements.

If combustible dusts are processed, the explosive characteristics need to be defined; these include

Type of dust
Dust particle size and size distribution
Explosibility index—explosion severity and ignition sensitivity
NEC Group E, F, or G

The quantity of dust involved, propensity of building design to collect dust, dust housekeeping, and dust removal systems also need to be evaluated.

PLANT LAYOUT: PHYSICAL FACILITIES

The type and spacing of process vessels, the equipment piping, and, particularly, the location and spacing of control and electrical equipment buildings are of special importance with respect to electrical and control system safety. Wherever flammable or combustible materials are involved, outdoor, open-air facilities are always preferred for fire and explosion reasons. Natural outdoor ventilation will dissipate any leaks or spills and minimize the probability of an explosive concentration from accumulating.

In warm or temperate climates, the open-air facility is a way of life, but in colder climates enclosed or partially enclosed structures are found for operational reasons. Compressors and similar equipment that have a high frequency of leaks should not be enclosed in small buildings; there have been explosions in compressor buildings.

Control and electrical equipment rooms or buildings should not be located in process units if at all possible. If they are located in or close to process equipment that contains flammable or combustible materials, then pressurized ventilation systems must be provided in compliance with NFPA 496.

Factors to consider in evaluating plant layout and physical facilities include the following:

- Outdoor or enclosed process facilities.
- Distance between process equipment and control buildings and electrical rooms or buildings.
- Presence of tall towers or structures that provide lightning protection.
- Age of equipment.

- Equipment that may require protection from vibration.
- Location of power feeders and electrical cable trays or conduit banks: Are they subject to damage from moving equipment chemical spills or fires?

3

Hazardous (Classified) Locations: Area Classification

The processing and storage of flammable gases and vapors are essential to the operation of most chemical processes. If these flammables leak or escape from piping, vessels, or other parts of the process, they can be ignited by any sparks or hot surfaces that may be in the flammable air mixture. The resulting fire and explosion can be catastrophic.

The Flixborough blast is a classic example, and the 1989 explosion at the Phillips Petroleum plastics plant at Pasadena, Texas, is the latest. The Phillips blast killed 22 people, and the blast wave was felt as far away as Houston. The safety stakes are high when flammables are present. Even small flash fires can result in severe burns and possibly death to operators.

One possible ignition source is electrical equipment or wiring. Sparks from relay contacts, wiring failures, switch contacts, hot surfaces of motors, lights, heaters, and even the hot filament of a broken flashlight bulb can cause ignition.

NATIONAL ELECTRICAL CODE TERMINOLOGY

Prevention of ignition from electrical equipment and wiring systems is the purpose of article 500, "Hazardous (Classified Locations," of the *NEC* (1990). This article includes the requirements for electrical equipment and wiring in locations in which fire and explosion risks exist because of the possible presence of flammable gases or vapors (Class I), combustible dusts (Class II), or ignitable fibers and flyings (Class III).

Safety of electrical facilities in these types of locations is a two-step process. First, the location must be classified with respect to the characteristics of the flammables and the probability that flammable concentrations with air may be present. This is called area classification, and it requires an analysis of the flammable characteristics of the particular gas, vapor, or liquid involved, the probability of it escaping from the process, and the physical environment at and around the release point. After a location is classified,[1] various protective measures must be applied to electrical facilities to prevent ignition, as described in the *NEC*.

Article 500 has some limitations. It only applies to flammable air mixtures and does not apply to oxygen-enriched or other mixtures. Nor does it apply to pyrophoric materials because they ignite spontaneously upon contact with air (and therefore electrical equipment would not be an ignition source). Nor is it concerned with nonelectrical ignition sources or protection against ignition by static electricity of lightning discharges.

NEC Articles 500-2 through 500-7 (1990) provide the basic definitions and concepts for classifying locations. According to the articles, Class I locations are those in which flammable gases or vapors are or may be present in flammable concentrations. Flame propagation occurs in a range of volumetric concentration that varies with each flammable gas or vapor.

Class I locations are further subdivided into two risk classes: Division 1 and Division 2. Article 500 indicates that Division 1 locations are where flammable concentrations can be present under normal operating conditions or exist frequently because of repairs or maintenance operations or leakage and where breakdown or faulty operation of the process equipment might release flammable concentrations and simultaneously cause an electrical failure, which could provide a source of ignition. The articles also provides the following examples of Division 1 locations:

- Locations where flammables are transferred from one container to another
- Interiors of spray booths
- Open tanks or vats
- Inadequately ventilated pump rooms
- Other locations where flammables are likely to occur in normal operations

Clearly, open process equipment and inadequate ventilation are conditions that result in Division 1 classification.

---

[1]The term *classified* is used to replace *hazardous* because it is specific to this article whereas *hazardous* is a general term.

Division 2 locations are indicated as any of the following:

- Locations where flammables are contained in closed piping or vessel systems for which leaks or releases can occur only due to accidental breakdown, rupture, or abnormal operation of equipment
- Locations where flammable concentrations are prevented by positive mechanical ventilation, which could become flammable if the ventilation fails
- Locations adjacent to Division 1 locations where transmission of gases or vapors can occur but is prevented by positive pressure ventilation

Piping systems that do not contain valves, meters, or similar items would not ordinarily require classification even though they contain flammable gases or vapors, nor would areas where flammables are stored in sealed containers.

Determining the existence and extent of Division 1 and Division 2 locations is the most difficult and controversial task associated with classifying a location. The definitions in Article 500 are in broad and general terms and therefore somewhat difficult to apply to actual process situations. Fortunately, there are other sections of Article 500 as well as NFPA and American Petroleum Institute (API) standards that provide guidance in classifying locations.

In addition to class and division, it is essential to define the *NEC* group and T number to define the classification of a location completely. The explosion characteristics of gases and vapors vary considerably, depending upon the material involved. Those related to the approval of electrical enclosures for Class I locations and primarily explosionproof equipment (explosionproof equipment contains the explosion inside the enclosure and does not allow it to propagate outside the enclosure) are

- Maximum explosion pressure inside the enclosure
- Maximum safe clearance or gap between surfaces of a cover joint
- Autoignition temperature

Flammable gases and vapors with similar characteristics are grouped together for convenience in testing, listing, and approving equipment. Autoignition temperature is independent of pressure and safe gap and is covered by T ratings.

Typical materials in the group classifications are as follows:

Group A Atmospheres containing acetylene
Group B Atmospheres containing hydrogen fuel and combustible process gases containing more than 30% hydrogen by volume, butadiene, ethylene oxide, propylene oxide, acrolein, and gases of similar explosive characteristics
Group C Atmospheres such as cyclopropane, ethyl ether, and ethylene
Group D Atmospheres such as acetone, ammonia, benzene, butane, ethanol, gasoline, hexane, methanol, and methane

A listing of more than 200 flammable gases, vapors, and liquids with their *NEC* groups and autoignition temperatures is provided in NFPA Standard 497M (1986b). This document can be used to determine the *NEC* group and autoignition temperature. The autoignition temperature of a sample of the actual process material can also be determined by test. Also, the *NEC* group can be assigned after explosion pressure and safe-gap measurements are made or can be estimated based on the chemical structure, flame temperature, and other properties as indicated by Chamlee and Woinsky (1974). About 36 chemicals have been tested and classed into a group, but many of the listings in 497M have not been tested.

The Underwriters Laboratories, Inc. (UL) test apparatus used to measure explosive pressure and safe gap is called the Westerberg apparatus. It consists of a test vessel with two chambers separated by an adjustable joint simulating the opening on the cover of an explosionproof (EP) enclosure. Ignition is initiated in the test gas in one chamber, and the joint is adjusted to determine the largest opening that will quench the flame and not permit ignition of the same gas in the adjacent chamber. The opening, or maximum experimental safe gap (MESG), is the principal determinant of group classification. Examples of typical measured values are as follows:

Group	MESG (in.)	Maximum Explosion Pressure (PSIG)
A	0.003	11,140
B	0.003	845
C	0.027	200
D	0.029	160

Note that a Group D enclosure installed in a location classified because of the presence of hydrogen may have too large of a gap to prevent flame propagation and may not have adequate mechanical strength.

Article 500 requires that electrical equipment be approved not only for the class but also for the group and T number. It is essential that heat-generating electrical equipment, such as lighting fixtures, motors, solenoids, and heaters, have heated surfaces that are temperature limited to below 80% of the autoignition temperature of the particular gas or vapor that may be present. In Division 2 locations, this applies to normal operation; in Division 1 locations, it applies to normal and abnormal operation such as a stalled motor.

T numbers are used to indicate safe surface temperatures for heat-generating equipment. In other words, they indicate the maximum surface temperature of the equipment. T numbers are defined in Table 500-3(b) of the *NEC*. Equipment must be selected with a T number that does not exceed the autoignition temperature of the particular gas or vapor. The hottest equipment is indicated by T1, 460 °C; the coolest is T6, 85 °C. If a gas has an autoignition temperature of 200 °C, heat-generating equipment having a T3A, 180 °C, or lower T rating must be used. Use of a higher temperature rating (i.e., lower T rating) could be a continual ignition source.

Classification therefore consists of the following definitions:

Class Class I flammable vapors or gases
Division Division 1 or 2 or unclassified
NEC Group Group A, B, C, or D
T number T1 through T6

Defining the existence and extent of Division 1 or 2 zones is by far the most complicated and controversial aspect of classification. Class, group, and T number can be derived by testing or by referring to the NFPA 497M (1986) or other standards.

FLAMMABILITY CHARACTERISTICS

The first step in classification is characterizing the particular gas or vapor with respect to its flammable characteristics. The characteristics needed to apply the classification figures found in the NFPA and API classification standards are

State gas or liquid
Explosive limits in percent by volume
Vapor density as a gas or as vapors of a liquid
Flash point for liquids

The other flammability characteristics relevant to approval of electrical equipment and installations practices are

NEC Group A, B, C, or D (for Class I)
T number
Ignition energy relates to approval of intrinsically safe or nonincendive systems

Materials classified as gases in NFPA 325M (1984) have the highest flammability rating listed.

Gas density influences the path of a release or leak. Gases that are lighter than air, such as hydrogen, tend to rise and dissipate quickly and will not produce significant concentrations at grade level. In fact, unless restricted by roofs or other enclosures, lighter-than-air gases seldom form flammable mixtures at grade level.

Heavier-than-air gases tend to accumulate along grade level. If there is very little breeze or wind, dispersion may be slow and the gas release may remain at grade unless it is moved by natural and forced ventilation. Gases more than 50% the density of air tend to act like heavier-than-air gases.

Gas temperature will influence the dispersion process. Heavier-than-air gases at elevated temperatures tend to act like lighter-than-air gases until they cool; conversely, lighter-than-air gases, if cooled sufficiently, may act like heavier-than-air gases for some distance until they absorb heat.

Flammable limits influence the extent of Division 1 and 2 zones. The concentration of a flammable gas must reach a certain minimum volumetric concentration, referred to as the lower explosive limit (LEL), for ignition and flame propagation to occur. Below that concentration, the gas mixture is too lean. At a higher concentration, the upper explosive limit (UEL), the concentration is too high or too rich. In each case, measured values vary with the particular gas and with the test gas temperature and pressure. An increase of temperature or pressure will lower the LEL and raise the UEL. Flame speeds and explosion pressures peak within the flammability range while ignition energy reaches a minimum value.

Classification codes and standards treat hydrogen, lighter-than-air gases, heavier-than-air gases, and other gas groups separately and provide classification figures for each group.

Liquids, in general, do not burn; but if they are heated above their flash point, they produce flammable vapors. The flash point is the lowest temperature at which a liquid produces sufficient vapor to form an ignitable mixture with air at the surface of the liquid to ignite and propagate flame. Unless a liquid is at or above its flash point, it is not flammable. Flash point is measured by the closed cup or open cup American Society for Testing Materials (ASTM) standards. The closed cup flash point is usually lower than the open cup one. Measured values as indicated in NFPA 325M can be used to determine if classification is required. Flash points for various liquids include the following:

	Flash Point	
	°F	°C
Acetic acid	103	39
Acetone	−4	−20
Benzene	12	−11
Cyclohexane	−4	−20
Vinyl toluene	120	49
Nitrobenzene	190	88

Liquids with flash points less than 100 °F are defined as flammable. They can be heated by ambient air to above 100 °F. In some especially hot locations with a high exposure to sunlight, liquids with flash points less than 140 °F could be heated above their flash points and require classification. Liquids with flash points above 100 °F are defined as combustible and are only flammable if processed above their flash points. Liquids with flash points above 200 °F are not normally classified and are not listed in NFPA 497M.

The lower the flash point, the higher the risk. Acetone, for instance, is flammable in almost any situation, whereas nitrobenzene must be heated above 190 °F.

LEAK AND RELEASE SOURCES

The presence of flammable gases, vapors, or liquids in process piping or vessels does not imply that classification is required. For classification to occur, there must be a means for leakage or release outside the system. *NEC* Article 500-5 indicates that piping that contains flammables and is without meters, valves, or similar devices would not normally be considered hazardous.

Flammables can escape from the process by leakage at pump or agitator seals, control valve packing, or other similar leakage points or at open process equipment like drum filling, filter presses, and other similar open operations. These leakage and open process conditions should be considered sources for the purposes of classification.

Releases can also occur due to catastrophic rupture or failure of piping or vessels. These major vapor cloud releases can cover large areas but are usually detected and stopped in a short time, typically in less than 15 minutes. Most of the time they dissipate into the air. Division 1 or 2 classifications are not appropriate for these situations because of the short duration (for most industrial electrical enclosures it would require over an hour for gas to diffuse inside the enclosure). The presence of other open ignition sources, such as open flames and automobiles, can become an ignition source.

In summary, classification should occur around sources of release. The shape and size of the classified zones depend on the flammable, ventilation, and process conditions.

VENTILATION

Air movement is effective in dissipating leaks and releases of flammable gases and vapors. Any breeze or air circulation can dissipate the release. If a portable combustible gas analyzer is used to test for combustible mixtures at open, unheated process equipment in an outdoor location on a breezy day, a flammable concentration will not be indicated until the sensing probe is moved to within inches of the opening emitting flammable vapors.

Wherever possible, chemical process equipment and piping should be located outdoors, without roofing or walls. That is because any obstruction to airflow reduces the dissipation of leaks.

Temperature differences within a multilevel process structure generated by heated process equipment and piping can generate significant thermal drafts in a structure. Even a structure that is roofed can be considered adequately ventilated if it has no walls or obstructions.

Enclosed or partially enclosed process or equipment structures can be provided with mechanical ventilation consisting of fans or blowers and the associated duct work and controls. The amount of ventilation should be such that flammable concentration does not exceed 25% of the LEL. This criterion can be applied if the leakage or spill rate is known. In the absence of such data, NFPA 30 indicates that 1 cubic foot per minute per square foot of surface area is adequate for declassification. For a level that is 10 feet high, this would require six air changes per hour. The air intake to the ventilation system must be in a location that is not in or close to a classified location. In a large chemical complex with process units close to one another, it may be difficult to find "safe" air and may require that the air pickup point be at a higher location than the ventilated area. The ventilation should cover all parts of the process unit that are classified. Mechanical ventilation must be protected against failures. Loss of ventilation should be alarmed by an airflow switch or an equivalent sensor in a control room, and corrective action should be taken quickly.

Division 1 zones can be reduced to Division 2, if ventilation systems that provide adequate ventilation and are alarmed in the event of a failure are provided, Division 2 zones can be reduced to declassified.

PROCESS CONDITIONS

Process conditions influence classification. For a given flammable, leak size and degree of ventilation, process temperature and pressure, and to a lesser degree flow rate and equipment size influence the size of the flammable zone. Unless a liquid is at a temperature above its flash point, either by ambient air or process heating, it will not produce flammable vapors if a leak occurs. The higher the pressure, the greater the leakage. High-pressure processes should have larger classified locations.

VAPOR BARRIERS

Walls without windows or doors or any other potential openings or leakage points limit the dispersion of heavier-than-air gases or vapors along the ground and therefore limit classified locations. The classification figures in NFPA 497A and API RP500 show that unpierced walls limit classification, but classification extends outside of a pierced wall. If a building has unpierced walls extending into a Division 2 location, the inside of the building is unclassified. Unpierced walls must provide a total barrier to gas or liquid to be effective in limiting classification. Any openings for pipe or conduit must be sealed and remain sealed.

NONELECTRICAL IGNITION SOURCES

Open flames at burners, fired heaters, hot surfaces of pipes, or vessels at a temperature above the autoignition temperature of any flammable gas that may be present are continuous nonelectrical ignition sources. Since the purpose of classification is to eliminate ignition sources, electrical classification in the immediate area of this equipment is inappropriate. Prevention against ignition is achieved by eliminating the classified area or removing the source. Even if the open flame or hot surface is outside a Division 2 classification, in the event of a large vapor release, it can and has provided a source of ignition. If the velocity of combustion air into a burner is faster than any possible flame speed, ignition may not occur.

ELECTRICAL CLASSIFICATION PRACTICE

An editorial in the July 1989 issue of *Control Magazine,* "Hazardous Area Classification : Ambiguity Still Exists," accurately describes the state of technology in electrical classification. It includes some of the following points:

1. Overclassification is the general rule.
2. Misunderstanding of code terminology is a reason for overclassification.
3. Added equipment costs from overclassification are significant.
4. Word definitions are difficult to apply.
5. Manpower limits the application of NFPA and API guidelines.

It is difficult to understand why industry spends so much time, effort, and money on explosionproof equipment and other electrical requirements for classified locations and so little effort on classifying locations to determine if it is needed or, more important, that it is the right type (i.e., correct group and T number).

The classification of locations is a critical safety issue and deserves to be treated as an important engineering tasks. Examples of improper classification follow:

1. Classification at open flames, roads, and other locations where nonelectrical continuous ignition sources are present.
2. Classifications without indicating the *NEC* groups (A, B, C, or D) or T numbers or indicating the wrong groups.
3. Blanket classification of an entire block or unit without consideration of the variables necessary to classify. In one unit, an entire building was blanket classified except for the roof, which was unclassified. The roof has open vents that could breathe flammables. This is an example of not considering leak sources.

   In another operation, a fire and explosion occurred just outside a pierced wall of a building. An electrical panel, was in close proximity. The panel was not considered to be in a classified location because the building was blanket classified without consideration of a pierced wall. (NFPA 497A indicates that classification is required outside of a pierced wall.)

   In another operation that processed hydrogen, the entire unit was blanket classified, Class I, Division 2, Group B (hydrogen). Hydrogen gas has a density of 0.1; therefore, hydrogen classifications do not normally extend any significant distance at grade. During the design phase of the project, it was difficult to find Group B equipment, so the classification was incorrectly changed to Group C.

The price for blanket classifying, that is, not following an engineering approach and not using NFPA and API standards, is severe and includes the following.

Safety

1. Locations are not classified that should be. This is primarily because of not classifying around sources.
2. Failure to recognize methods to reduce or eliminate classified locations.
3. Wrong group or no T number.
4. Compromise reliability of electrical equipment enclosures. Explosion-proof enclosures may have to be used instead of watertight or corrosion-resistant enclosures in locations that are overclassified.
5. Use of conduit instead of cable tray. Conduit can pipe process fluids into control and electrical rooms and enclosures.

Costs

1. Additional equipment costs for explosionproof and other types of approved equipment
2. Additional design, specification, and drawing costs associated with classified locations
3. Additional installation costs of classified location installations, for example, explosion seals
4. Additional checkout and inspection costs to verify that the electrical facility complies with Article 501
5. Additional maintenance costs associated with explosionproof and other special equipment and facilities required for classified locations

In addition to the above, in the event of an accident or audit, OSHA or other regulatory agency citations and fines are possible as a result of improper classification installations.

ANALYTICAL APPROACH TO CLASSIFICATION

The analytical approach to classification includes the analysis of the factors that influence classification coupled with sound engineering judgment and the development of area classification drawings. The factors that influence classification are as follows:

1. The flammable properties of the particular material:
   State Gas or liquid
   Explosive limits
   Vapor density
   Flash point for liquids
   NEC Group A, B, C, D
   T number based on autoignition temperature
2. Process conditions Primarily temperature and pressure and to a lesser degree flow and volume
3. Ventilation Outdoor locations—open construction
   Enclosed facilities
   Indoor location, with or without mechanical ventilation

With this data, using area classification national standards, classification drawings should be developed and updated as process or equipment changes occur.

The following additional factors should be considered in applying the classification figures in API RP500 and NFPA 497A. See Figures 3-1 and 3-2 from API RP500A (1982) and RP500C (1984).

1. Liquids do not require classification unless they are at a temperature at or above their flash point. If the flash point is below 100 °F (140 °F

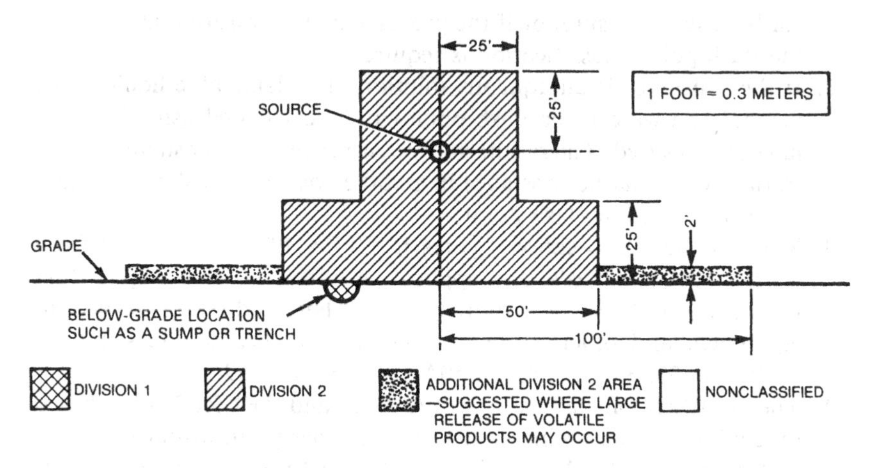

FIGURE 3-1. Example of the area classification figures in NFPA 497A and API RP500. Note the large distances associated with flammable gases. (American Petroleum Institute, API RP500A, 8, Fig. 2.)

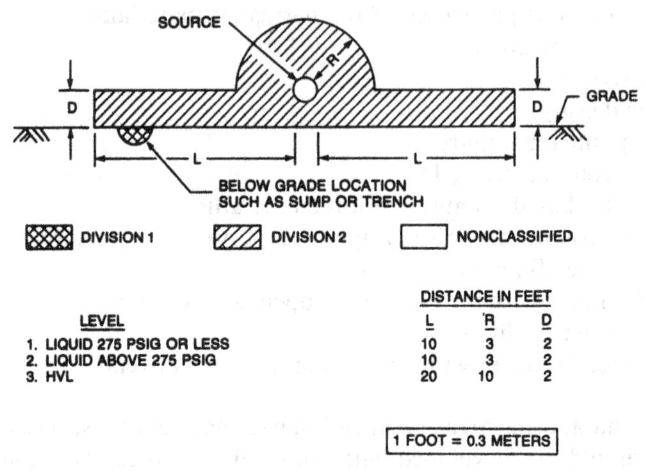

NOTE:
Shown for piping installed above grade. Also applies to buried piping with above grade vents, such as a motor operated valve with the valve body buried and the motor operator above grade.

FIGURE 3-2. Example of the area classification figures in NFPA 497A and API RP500. Note the increase in the size of the classified area with high pressure and highly volatile liquids. (American Petroleum Institute, API RP500C, 9, Fig. 4)

for high sun exposure) or if the processing temperature is at or above the flash point, classification is required.

2. If the substance is an aqueous solution of a flammable liquid (e.g., alcohol and water), the flash point can be calculated using Raoult's laws or measured. The results may be surprising. A very small concentrations of a volatile substance in an otherwise high flash point liquid can result in a low flash point.
3. Most of the classification figures in NFPA 497A are derived from other NFPA standards, and application of these figures implies that the facility follows the requirements of these codes with respect to equipment and other characteristics. The applicable codes include NFPA 30, 33, 34, 35, 36, 45, 50A, 50B, 58, and 59A.
4. The classification figures in NFPA 497A and API RP500 are based on equipment and piping systems that are designed, installed, tested, and maintained for high reliability chemical process or refinery service. For these systems, leaks are the exception and ruptures or failures are a low probability, that is, a once-in-a-lifetime event. These systems should be designed, installed, inspected, tested, and maintained in compliance with the applicable American Society of Me-

chanical Engineers (ASME), API, and American National Standards Institute (ANSI) piping and vessel codes and practices. Failure to do so requires a reevaluation of classification and other safety and environmental practices. For the purpose of classification, the following parts of the process equipment and piping systems can be considered sources in the application of classification figures:

- Packing or seals on pumps, control valves or agitators.
- Hose connections.
- Drains and overflows.
- Vents and sampling points.
- Pressure relief devices.
- First piping flanged connection off a vessel if the process is at high pressure or subject to vibration or high temperatures or if the vessel is a compressor. Other piping flanged connections or manual valves would not normally be considered sources for the purpose of classification.
- Piping or process connections that require frequent maintenance due clean out of process material.

5. NFPA 497A and API RP500 indicate that the Division 1 or Division 2 zone extends from leakage sources. Where the leak sources are close together (e.g., in a batch reactor process), it is not practical to zone classify. A larger area could be classified, considering the density of sources. Where flammable gases are processed at high pressures in large petrochemical type processes, large classified zones are appropriate and are indicated in API RP500.
6. Locations without leak sources or open process equipment, or where flammable liquids are processed below their flash point, or where aqueous solutions are below their flash point may not require classification .

ELECTRICAL CLASSIFICATION VERSUS EPA AND TOXICITY REQUIREMENTS

Chemical-process facilities are required to reduce and limit emissions to meet EPA and state regulations. This effort requires eliminating or reducing open process equipment and leakage sources. The acceptable concentrations to meet these requirements are far below the concentrations that can produce explosive (i.e., the LEL's). Containment of process material in locations where field operators work and the process material presents a health hazard is also required. Some process units have analyzers that set off an alarm at concentrations in parts per million; corrective action to repair the leak follows. The LEL for flammable gases or vapors is in percentage or thousands of times higher than EPA or safe health exposure levels. Therefore, it would seem difficult to find any large classified areas in facilities that meet these requirements. Division 1 would be hard to find, if it exists at all, and Division 2 would be limited. Blanket classification of a unit as Division 1 or 2 seems inconsistent with meeting EPA and operator safe health exposure levels.

REDUCING CLASSIFIED LOCATIONS

A worthwhile goal of every chemical process facility is to eliminate or reduce classified locations, certainly Division 1 locations. To accomplish this goal,

1. Reduce or eliminate leak sources and open process equipment.
2. Change facility from indoor to outdoor. Do not provide small buildings around compressors or other equipment that processes flammables.
3. If the facility is in a building, provide adequate mechanical ventilation with safeguards.
4. Provide vapor barriers or unpierced walls to limit the extent of classified locations for heavier-than-air gases and vapors.
5. Promote ventilation throughout the structure.
6. Reduce process temperature to below the flash point for liquids.

The effect of these changes is indicated by comparing classification figures in NFPA 497A and API RP500.

DIVISION 3

Some companies use a concept called Division 3, which provides an extra margin of safety outside of Division 2. Division 3 is not an *NEC* concept; rather, it provides protection beyond the code. It applies to process situations where large releases of flammable gases or vapors could occur. Division 3 is a vapor cloud area covering large distances. The extent of the plume depends on the flammable gas or vapor, the volume released, the gas or vapor temperature at the release point, and the prevailing temperature and wind conditions. It is possible to calculate the shape and extent of the plume based on given conditions using gas dispersion technology.

Vapor cloud releases in a chemical-process facility are quickly detected and stopped within a short time, usually less than 15 minutes, which is considerably shorter than the time required for a gas or vapor to diffuse inside an enclosure. Industrial electrical enclosures that are enclosed and gasketed provide at least an hour diffusion time before the concentration of the gas or vapor inside the enclosure will reach the LEL. Electrical enclosures in outdoor chemical-process installations are required at least to be enclosed and gasketed in order for the internal components and wiring to survive in outdoor, wet, or hose down environment conditions; therefore, the enclosure provides Division 3, vapor delay protection without special effort. The Division 3 concept recognizes the need to provide enclosed and gasketed requirements outside of Division 2, where vapor cloud releases are possible. Not all processes are capable of large vapor cloud releases, but when they are, Division 3 provides added protection.

Probability Concepts: Division 0

The basic and fundamental nature of electrical classification and *NEC* protection against ignition is probabilistic. The probability of an ignition (PI) is the product of the probability that a flammable gas or vapor will be present at a concentration within the flammable range (PF) and, at the same time, the probability that an electrical ignition source (PE) (either heated electrical equipment above the autoignition temperature or an electrical spark with sufficient energy and intensity to cause ignition) occurs. That is,

$$PI = PF \times PE$$

It is a fundamental safety principle that no single failure should cause ignition. All the factors that should occur for ignition to take place are reduced to low probabilities, so the overall probability of ignition is essentially zero. Note that if a hot surface operates above the autoignition temperature, PE equals 1 and we are totally dependent on PF to prevent ignition.

Probability definitions of Division 1 and 2 have been proposed based on the percentage of time a flammable is at a concentration within the flammability range. A location is considered Division 1 if the time is more than 0.1% or approximately 10 hours per year and is considered Division 2 if it exceeds 0.01% or 1 hour per year up to 10 hours per year. The concept is useful in evaluating certain situations. For instance, if a process stream is only flammable during upset conditions that occur for 1 hour a year, the location could not be considered Division 1 and at the worst could be Division 2.

The *NEC* indicates that Division 1 are locations where the concentrations of flammables could be flammable under normal operations. But what about those situations where flammables are present a very high percentage of the time or continuously? They include the area above a floating roof in a storage tank or in a ventilation duct that is removing flammable vapors.

These unusual situations deserve special attention. They are special, high-risk situations referred to as Division 0, or Zone 0) in European terminology.

Zone 0 has been used in Europe and is an international concept. Its locations require special precautions. That is, some Division 1 electrical installations may not be appropriate for Division 0. Furthermore, safety in special situations and uniformity with international standards are the arguments for Division 0, which is presently not in the *NEC,* although it has been proposed a number of times.

Using probability concepts, Division 0 can be viewed as a location where the concentration is flammable more than 1% of the time or 100 hours.

ELECTRICAL CLASSIFICATION: SPECIAL CASES

High Flash Point Liquids

High flash point liquids deserve special consideration. The National Fire Codes define liquids whose flash points are above 100 °F as combustible and those above 200 °F as Class III B combustible liquids. NFPA 497M does not include Class III B liquids in the listings of classified gases and liquids. These liquids have a very low probability of ignition by properly installed and maintained general-purpose electrical equipment.

Certain heat transfer fluids have flash points above 250 °F and can form mists if released into the atmosphere. Although in the strictest sense these fluids do not require classification as Class I vapors and gases, precautions with electrical equipment and wiring must be taken because they can be ignited by an electrical spark. Mists act very much like dusts and require a significant amount of time, possibly a number of hours, before they could diffuse inside any enclosed electrical enclosure. These locations, where high flash point, Class III B combustible liquids are processed, should be treated as Division 3 locations, and NEMA 3, 4, or 12 enclosures should be used for any enclosures that have open contacts like relays, switches, or push buttons.

Protecting any open cable or wiring from these hot liquids is important. That is because it is possible for the liquid to cause the insulation on the wire or cable to melt and cause a spark, which can ignite flammable vapors.

Battery Installations

Battery installations for forklift trucks, uninterruptible power supplies (UPS), emergency lighting, and so on, do not require electrical classification because the volume of hydrogen gas that might be liberated is so small and the period of time (during the final stage of charging) is so short that almost any ventilation will eliminate the need for classification. Article 480-8(a) of the *NEC* requires that adequate ventilation be provided at battery locations to prevent the accumulation of hydrogen gas. Almost any natural ventilation is adequate. The ventilation requirements are in the range of one to two air changes per hour. Hydrogen gas can be evolved during the final stage of charging. The type of battery charger and batteries are factors in determining the volume of hydrogen evolved. The volume is so small, however, that only the immediate area around the vents is of any concern. There are precautions to observe whenever making connections at terminals near the battery vents. There are also precautions concerning handling batteries. Safety goggles are required, and other protective gloves and equipment must be used.

Process Control Buildings

Process control buildings typically include instrument control panels, process computer consoles and other computer equipment, alarms, recording and logging instrumentation, and other controls necessary for control room operations to monitor, control, and generally run the process. The buildings are also a gathering place for maintenance and other personnel, especially on hot days. Control buildings therefore represent a significant investment in instrumentation, engineering, and people, and the safety of these buildings is a very important issue.

Process control buildings should be located some distance from classified areas, if at all possible. Fifty feet is considered a minimum distance to allow for fire-fighting equipment. If a control building is located in a classified area, it is classified if any door, window, or other opening is within the classified zone.

The instrumentation and controls within a process control building cannot be made explosionproof, intrinsically safe, or nonincendive and are subject to frequent repair and adjustment. These control buildings must be purged and pressurized as well as ventilated and cooled because of the people and electronics in the building. NFPA 496, *Standard for Purged and Pressurized Enclosures for Electrical Equipment,* 1986 describes the requirements for pressurized control buildings or rooms in Class I locations. The requirements include the following:

1. The room or building shall be constructed to prevent the entry of flammable gases, vapors, or liquids.
2. The source of outside air for positive pressure ventilation must not

be exposed to flammable concentrations of gases or vapors, contaminants, dusts, humidity, toxic gases, and so on.

3. The ventilation duct system must be constructed of noncombustible materials, prevent the leakage of flammable vapors or gases, and be constructed to protect against mechanical damage and corrosion.
4. The positive pressure within the building must be maintained at a minimum of 0.1 inch of water with all openings closed and a minimum outward airflow of 60 feet per minute through all openings that can be opened: The pressure can drop below 0.1 inch of water column under this condition.

The positive pressure must not be so high that it is difficult to open doors. Four-tenths of an inch of water pressure is a reasonable design value.

The level of protection to ensure that the positive pressurization is above 0.1 inch of water is commensurate with the degree of declassification. Reduction of classification in the control building from Division 2 to unclassified (one level of declassification) is called Type Z purging and requires an alarm on loss of pressurization. A reduction in classification from Division 1 to Division 2 is called Type Y purging and requires an alarm since it is one level of declassification. A reduction in classification from Division 1 to unclassified (two levels of declassification) requires the removal of power to equipment that is not explosionproof or intrinsically safe.

Therefore, it is not practical to locate a control room in a Division 1 location. Process control buildings located in a Division 2 or 3 location must be constructed to prevent the entry of flammable gases, vapors, or liquids into the building. All possible ways should be considered. The control building outer walls should be vapor tight. Windows should be avoided. If they are used, they should not face facilities where flammables are processed or stored. Exit doors should also face away from flammables. Conduits, cables, instrument tubing, piping, and so on, should be examined as potential sources of process fluids that can and have entered control buildings.

Flammable vapors have entered control buildings through floor drains and during unusual, upset conditions and have backed up through toilet connections, causing control room explosions. Electrical conduit runs under the floor and through walls and have provided a channel for process gases and vapors to enter control rooms. Therefore, never have a direct connection through a conduit or cable from a process-connected instrument, such as, a thermocouple, resistance temperature detector (RTD), or pressure switch, into the control building. The thermowell or other process seal can fail, and the process fluid will pressurize the conduit or cable system, causing leakage into the building. This phenomenon is discussed in detail in Chapter 4.

Control building pressurization systems require a source of outside air that is totally free of flammable (or toxic) gases or vapors at all times. Even if the control building is located outside a Division 2 location, it may be in a Division 3 location where large and infrequent cloud or plume releases can occur. In a large chemical complex with many interacting process units, it may be difficult to find a location where large vapor releases cannot occur. In some situations, it has been necessary to install the air intake at the top of an adjacent building. Continuous combustible gas monitors at the air intake as well as in the building should be considered to protect against vapor or gas entry into a control building.

CONCLUSION

Area classification is a critical safety issue and is required by the OSHA regulations for the process safety management of highly hazardous chemicals. Safety against possible ignition and explosions of flammable materials is achieved in two steps. First, classify the location; second, provide the protective measures described in the *NEC* Article 500.

Area classification is not done very well in industry. Overclassification, usually blanket classification, is the rule, but underclassification is also possible. To help alleviate that situation, this chapter provided an analytical approach to classification based on sound engineering principles. The rational approach begins by considering the properties of the flammable (gas or liquid, flash point, etc.), the process conditions (temperature and pressure), and the environment (outdoors and ventilation).

Classification occurs around risk or leak points, which is a fundamental difference between blanket classification and analytical classification. The classification figures in the API and NFPA standards are a valuable resource in developing area classification drawings. Do not be preoccupied with developing detailed drawings using these figures; rather, develop sound drawings based on the application of the figures and good judgment.

Division 3 as discussed in this chapter is also a useful concept and should be applied to process situations where vapor cloud releases are possible so as to prevent catastrophic explosions.

Reducing or eliminating classified locations by providing ventilation, location outdoors, and vapor barriers; eliminating leaks; and reducing process temperatures and pressure should always be considered.

REFERENCES

American Petroleum Institute. January 1982. *API RP500A. API Recommended Practice, Classification of Locations for Electrical Installations in Petroleum Refineries,* 4th ed.

American Petroleum Institute. July 1984. *API RP500C. API Recommended Practice, Classification of Locations for Electrical Installations at Pipeline Transportation Facilities,* 2d ed.

American Petroleum Institute. October 1987. *APRI RP500B. API Recommended Practice for Classification of Locations for Electrical Installations at Drilling Rigs and Production Facilities on Land and on Marine Fixed and Mobile Platforms,* 3d ed.

Buschart, R.J. 1975. An Analytical Approach to Electrical Area Classification: Flammable Vapors and Gases. ISA Paper 75-763. *Advances in Instrumentation* 30(3):763.

Chamlee, R.D., and S.G. Woinsky. 1974. Predicting Flammable Material Classifications for the Selection of Electrical Equipment. *IEEE Transactions on Industry Applications* IA-10 (2, March/April):288–298.

Hunt, G.O. Outside Division Two. *Electrical Safety Practices* ISA Monograph 1972 (113):93–96.

National Fire Protection Association. 1984. *NFPA 325M. Fire Hazard Properties of Flammable Liquids, Gases, and Volatile Solids.*

National Fire Protection Association. 1986a. *NFPA 497A. Classification of Class I Hazardous (Classified) Locations for Electrical Installations in Chemical Process Areas.*

National Fire Protection Association. 1986b. *NFPA 497M. Classification of Gases, Vapors, and Dusts for Electrical Equipment in Hazardous (Classified) Locations.*

National Fire Protection Association. 1990. *NFPA 70. The National Electrical Code.*

National Fire Protection Association. *NFPA 496. Standard for Purged and Pressurized Enclosures for Electrical Equipment.* 1986

4

Electrical Equipment in Class I Locations

Electrical equipment and wiring in locations classified as Class I, based on the likelihood that flammable concentrations of vapors or gases can occur, must be protected so they cannot cause ignition of the flammable gases or vapors. Ignition can occur from an arc, from a contact or wiring failure, or by the hot surface of a motor, lighting fixture, or heater. Protection of electrical equipment and wiring does not prevent ignition from other spark sources, for instance welding, static electricity, lightning, or cutting, or from other hot surfaces, for instance, steam lines, gas heaters, or open flames at burners. Other codes, standards, and safe working practices provide the safeguards for these hazards. The criteria for protection is found in Articles 501-1 through 501-17 of the *NEC.*

It is essential that the classification be defined. The Class (I), Division (1 or 2), Group (A, B, C, or D), and T number must be defined for the particular location using Article 500 and other NFPA standards as previously discussed.

Articles 501-1 through 501-17 cover the requirements for transformers, capacitors, meters, instruments, relays, wiring methods, switches, motor controllers, motors, lighting, heaters, flexible cords, signaling, alarm, remote control and communication systems, grounding, and surge protection. These sections of the code describe the special rules for electrical facilities in Class I locations in addition to the requirements in other sections of the code. In particular, the sections regarding working space around electrical equipment, guarding live parts, overcurrent protection, grounding, wiring materials and methods, motors, motor controls, lighting equipment, transformers, and so on, all apply to classified as well as unclassified areas and are essential to safety in these areas.

Grounding and overcurrent protection are especially important since they eliminate or minimize the time arcs or sparks can be present if electrical equipment or wiring fail.

Articles 501-1 through 501-17 recognizes several ways to protect against ignition of flammable gases or vapors. They include

1. Location outside the classified location
2. Explosionproof apparatus
3. Purging and pressurization of enclosures and buildings
4. Hermetically sealed electrical contacts
5. Oil immersion of contacts
6. Intrinsically safe systems for instrumentation and controls
7. Nonincendive systems for instrumentation and controls
8. Special wiring system requirements—explosion seals, grounding, and so on

It is important to remember that the above protective measures do not classify the locations; they are a response to the classification. Classification depends on the flammable, the process, and the environment. Some of the protective methods depend on the particular gas or vapor; therefore, the Group (A, B, C, or D) and the T number on maximum surface temperature must be known.

ELECTRICAL FACILITIES OUTSIDE CLASSIFIED LOCATIONS

Wherever possible, electrical equipment should be located outside of classified locations. Process instrumentation, motors, pipeline tracing, lighting, and other process-connected electrical equipment are an integral part of process piping and equipment and may of necessity be in classified locations. Wherever possible, however, electrical equipment such as control panels, power distribution, panels, control instrumentation, programmable logic controllers, motor starters, and disconnects should be located in process control and electrical rooms or buildings outside the classified location that would also provide protection from the chemical plant environment. Motor starters and disconnects, control panels, lighting equipment, and control power distribution panels located in a process unit may require costly explosionproof or purged and pressurized enclosures if they are in a Division 1 or 2 location. As a minimum, the enclosures must be protected against rain, dirt, dust, hose down, and chemical corrosion. This will require National Electrical Manufactures Association (NEMA) 3, 3R, 4, or 12 enclosures, which are more costly than control room enclosures. Maintenance in the process area exposes maintenance personnel and the inside of electrical enclosures to the process environment. Motor disconnects are sometimes located at the motor in the process area. This is not necessary since disconnects in motor control centers located in electrical rooms are acceptable.

NATIONAL ELECTRICAL CODE

The requirements for electrical equipment and wiring are indicated in Article 501 for Division 1 and Division 2 locations. In Division 1 locations, normal operation and failure of equipment and wiring are considered potential ignition sources. Division 1 requirements are considerably more restrictive than Division 2 because of the higher probability of a flammable vapor or gas occurring.

Any failure must be considered a possible ignition source. In Division 1 locations, equipment and wiring must be explosionproof and intrinsically safe, or the enclosures must purged and pressurized by Type-X or Type-Y purging.

In Division 2 locations, because the probability of a flammable mixture is lower, only equipment that can cause ignition under normal conditions needs to be protected. This means any ignition-capable contact device such as push buttons, limit switches, and starter contacts must be in a suitable enclosure. Explosionproof apparatus and intrinsically safe systems must be listed and labeled for the particular gas or vapor, the correct *NEC* Group (A, B, C, or D), and the correct T number, if the electrical equipment generates heat.

Equipment for hazardous locations is listed by accredited testing and certification laboratories, including Underwriters Laboratories, Factory Mutual Research Corporation (FM), ETL Testing Laboratories, MET Electrical Testing Laboratory, Inc., and other testing and certification agencies approved by OSHA for third-party certification.

The Underwriters Hazardous Location Equipment Directory (the Red Book) provides a listing by type of equipment, manufacture, and division and group. In Division 2 locations, contact devices such as push buttons or limit switches can be hermetically sealed, immersed in oil, or in nonincendive circuits. Electrical motors in Class I, Division 1 locations must be explosionproof, totally enclosed with positive pressure ventilation from a source of clean air, or totally enclosed inert gas filled and pressurized. Explosionproof motors are available for Groups C and D but not for Group B (hydrogen, etc.)

The surface temperature of an EP motor cannot exceed the ignition temperature of the vapor or gas under normal and abnormal conditions, including locked rotor and single phasing. This requires a thermostat for certain sizes and types of motors. This thermostat is interlocked to the motor control electrical circuit.

Explosionproof motors (see, e.g., Figure 4–1) are marked with a T number to indicate the maximum permissible surface temperature. They are not manufactured in larger sizes; typically, 100 HP is the limit. Purging and pressurization are used in larger sizes. In Division 2 locations, standard induction motors without contact devices are used. Direct current motors must be explosionproof or purged and pressurized because of the motor brushes. Most single-phase motors should be explosionproof because of the starting centrifugal switches.

Larger motors located outdoors are sometimes provided with condensation heaters. The heater surface temperature must not exceed 80% of the autoignition temperature of the particular gas or vapor. Motor fans must be made of nonsparking material.

Transformers, capacitors, and high-voltage starters usually cannot be supplied in explosionproof enclosures and cannot be located in Division 1 locations unless the enclosure is purged and pressurized or located in a pressurized electrical building. In Division 2, transformers need only com-

FIGURE 4–1. Explosionproof motor. Note the heavy cast-iron construction, especially at the terminal box on the side. The motor has a label indicating it is UL listed for Class I, Group C or D, and a T number. The motor is manufactured so openings at the terminal box cover and on the shaft are narrow enough to quench any flame resulting from an internal explosion. All bolts on the cover must be tight. Some EP motors have internal thermostats to deenergize the motor if the surface temperature is excessive. (Reliance Electric Company, Cleveland, Ohio.)

ply with other code requirements. Any switches on the transformer must comply with Division 2 requirements.

Dry type transformers must have surface temperatures less than 80% of the autoignition temperature of the particular gas or vapor.

Instrumentation in Division 1 must be explosion proof, purged, and pressurized or intrinsically safe. In Division 2, general-purpose enclosures are permitted if the contacts are hermetically sealed, immersed in oil, or in nonincendive circuits.

EXPLOSIONPROOF APPARATUS

Explosionproof apparatus is apparatus provided in an enclosure for electrical equipment and wiring that prevents an internal explosion of a particular gas or vapor from igniting external concentrations of the same gas or vapor. It must be strong enough to withstand an explosion of a particular gas within it (e.g., Figure 4–2); it must be flame tight so flame generated inside the enclosure that passes through covers and openings is quenched so it does not ignite the particular vapor outside; and it must be cool enough on any outside surface so it could not autoignite the external flammable gas or vapor.

Motors, generators, motor starters, disconnects, power panels, lighting fixtures, limit switches, instrument transmitters, push buttons, junction boxes, and other electrical enclosures can be supplied in explosionproof enclosures listed for Groups C and D, but not all equipment is available in Group B enclosures. If a surrounding flammable vapor or gas is at a high enough concentration for a long enough time so the flammable concentration within the enclosure reaches the lower flammable limit and an ignition source occurs from a spark or hot surface, the resulting flame and explosion will stay within the enclosure. Explosionproof enclosure testing and manufacturing is a mature business. Underwriters Laboratories have for years performed extensive testing and evaluation of this type of equipment and have documented the results in a number of reports.

Based on these reports, UL, FM, and other standards require explosionproof enclosures to confine the explosion with significant margins in safety. For instance, UL Standard 514 states that an explosionproof enclosure must be capable of withstanding a hydrostatic pressure of four times the maximum pressure of an internal explosion of a specific gas or vapor without rupturing or deforming. To determine compliance, UL requires enclosure strength tests or calculations. For cast metal, it requires a 1-minute hydrostatic pressure test of four times the peak explosion pressure measured during explosion tests. If calculations are used instead of tests, the design safety factor is five instead of four.

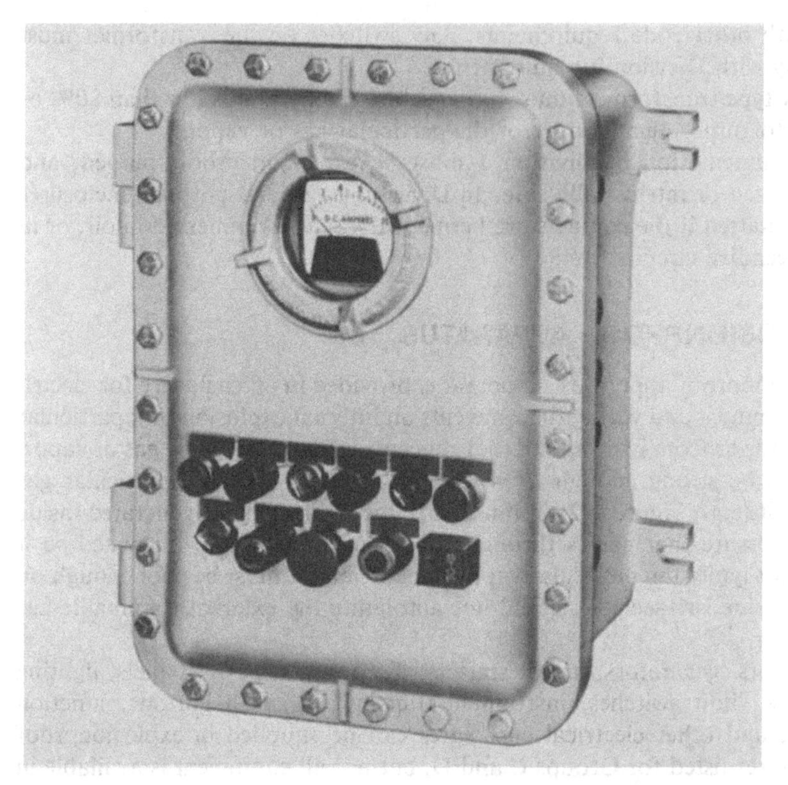

FIGURE 4-2. Explosionproof control box. The enclosure is designed to contain an explosion that might occur inside the box. The ammeter window and motor controls on the cover are designed to maintain the explosionproof integrity of the enclosure. The bolts on the cover must be tight. (Adalet-PLM Company, Cleveland, Ohio.)

Joints on covers, shafts, and other openings in explosionproof apparatuses must be narrow enough and long enough so the combined effect quenches any internally generated explosion and flame front and prevents ignition of the flammable gas or vapor on the outside.

There are two types of cover designs: ground joint and threaded. The ground joint refers to the smooth surfaces that close together on the covers. The mating surfaces on the cover and enclosure are machined to a fine finish. Gap and flame path vary considerably based on a number of factors. A longer path permits a wider gap.

A threaded enclosure cover must have a minimum of five threads engaged, even though laboratory tests on Group C and D enclosures indicate that a one and one-fourth thread engagement will quench the flame.

Underwriters Laboratories have done considerable testing under varying conditions on various types of enclosures and materials. The empirical data provide the basis for enclosure design with a significant margin of safety. Flame propagation resulting from the reaction of oxygen in air and a flammable gas or vapor in the presence of an ignition source occur only when the volumetric concentration is within certain limits, which are different for each gas or vapor. Outside these limits, localized burning occurs at the source of ignition, but burning ceases upon removal of the source of ignition.

A concentration exists between lower and upper limits that will produce the most violent explosion for a given gas or vapor. The concentration that produces maximum flame propagation is different from that which produces the highest pressure. Flame propagation is relatively slow at concentrations near the flammable limits and peaks somewhere between. Explosion pressure acts in the same manner.

During the testing of explosionproof motors, the measured pressure and rate of pressure rise are significantly higher if the motor is running due to turbulent effects of the internal air.

The volume and shape of the enclosure also influence explosion pressure; that is, larger volumes producing higher pressures. The shape and separation of compartments within an enclosure can result in ignition in one compartment, causing abnormally high pressures in another due to a phenomena called pressure piling. This occurs as a result of compression of the unburned mixture in the flame front. If the gas–air mixture in part of an enclosure is ignited, the exploding mixture will expand, which compresses the unexploded mixture, which in turn is ignited. The process is cumulative, raising the resulting pressure. Limiting the volume reduces the pressure rise in enclosures as well as conduit systems. Pressure piling between enclosures is limited by explosion seals in conduit or cables.

Wiring methods in Division 1 locations include conduit and type mineral insulated (MI) cable. All boxes and fittings shall be explosionproof. Threaded joints on conduit and enclosures should be connected with at least five threads fully engaged with a 3/4-inch-per foot taper. Flexible connections should be explosionproof braided metal, usually limited to 36 inches.

Receptacles and plugs will be explosionproof in Division 1 and 2 locations. Lighting fixtures must be explosionproof in Division 1, but enclosed and gasketed fixtures are permitted in Division 2 (see Figures 4–3, 4–4, and 4–5). Wiring methods in Division 2 locations include cable tray with various types of cable including tray cable (type TC cable).

SEALING PROCESS-CONNECTED INSTRUMENTATION

Process-connected instruments, for instance, thermocouples pressure switches, can fail and cause the transmission of toxic and/or flammable gases or

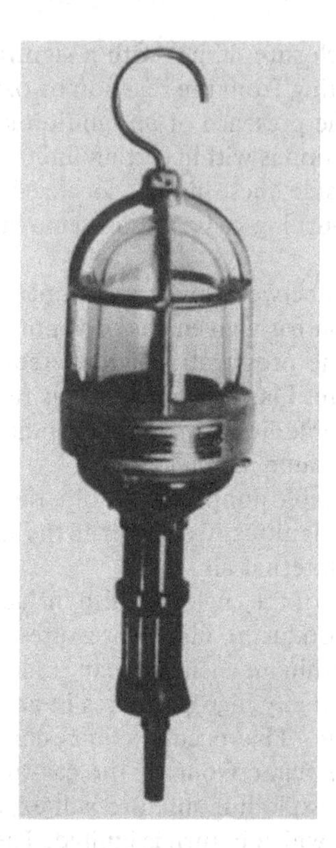

FIGURE 4-3. Explosionproof portable lighting fixture: a Crouse-Hinds type EVH explosion-proof portable light. These types of fixtures are required in a Class I, Division 1 or 2 location. The flexible cords connected to these lights must be the extra hard usage type and include a grounding conductor. Replacement of the globe should follow the manufacturer's recommendations. (Crouse-Hinds Electrical Construction Materials Company, Syracuse, N.Y.)

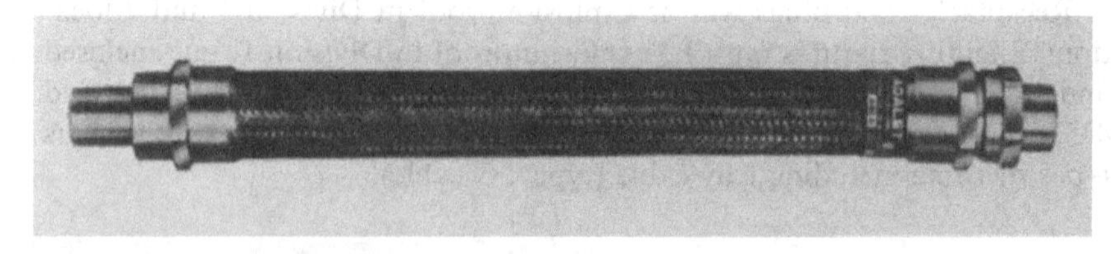

FIGURE 4-4. Explosionproof flexible conduit. This type of flexible connection is required in a Class I, Division 1 location, where a flexible connection to fixed devices is required. The conduit is constructed of braided metal and is labeled for the class and group. (Adalet-PLM Company, Cleveland, Ohio.)

FIGURE 4-5. Factory-sealed push button: N2S motor control station factory sealed for a Class I, Division 2 location. This type of control station does not require an external explosion seal and is watertight and dust tight. Explosion seals are field-installed concrete barriers in conduits and cables intended to block the passage of flames in an explosionproof enclosure. They are required within 18 inches of an arcing device and in other situations as described in the *NEC*. Explosion seals are difficult to install and should be avoided when possible by using factory-sealed devices. (Crouse-Hinds Electrical Construction Materials Company, Syracuse, N.Y.)

liquids into an electrical or process control room. This has happened a number of times and the results are serious. In one case, a thermowell ruptured due to corrosion or damage from an agitator blade; the thermocouple head was pressurized with the process fluid, including the open end of the thermocouple cable. After a period of time, the fluid, which is highly toxic, was forced by capillary action through the interstices of the cable insulation between conductors all the way through a considerable length of cable and over a time period of a number of hours into an instrument cable cabinet in the control room. Fortunately, the process fluid has a pungent odor and was detected by an operator before the concentration reached the lower flammable limit.

It is a fortunate circumstance of nature that most flammable materials have an odor at concentrations considerably below the lower flammable limit and therefore are detectable long before they can explode. Unfortunately, it is not true of all flammables.

In another situation at a liquified natural gas (LNG) facility, a canned pump in air cryogenic service experienced a leak at the motor leads that allowed process fluid to fill the motor lead terminal box, pressurize the box, and travel through an explosion seal and 200 feet of underground conduit into a switch room where the vapors were ignited by air arcing contact. The switch room was destroyed and one man killed.

Note that the process fluid passed through an explosion seal. Explosion seals prevent propagation of flames or pressure waves but are not intended to be, nor will they be, gastight or watertight (see Figure 4–6). A properly installed explosion seal will leak a small amount of pressurized fluid at one end. Conduit systems provide a pipeway for any process fluid. Transmission through the interstices of cables is possible, but the leakage rate is dependent on the size of the cable, length of the cable, process fluid and pressure, and time. Explosion seals do not prevent the propagation of process fluids.

A rule to remember is never provide a direct connection from a process-connected instrument or canned pump into a process control room or electrical room or enclosure either through conduit or cable. To do so involves a serious safety risk and violation of the *NEC*.

Article 501-5, section (f), Sealing and Drainage, indicates that to prevent

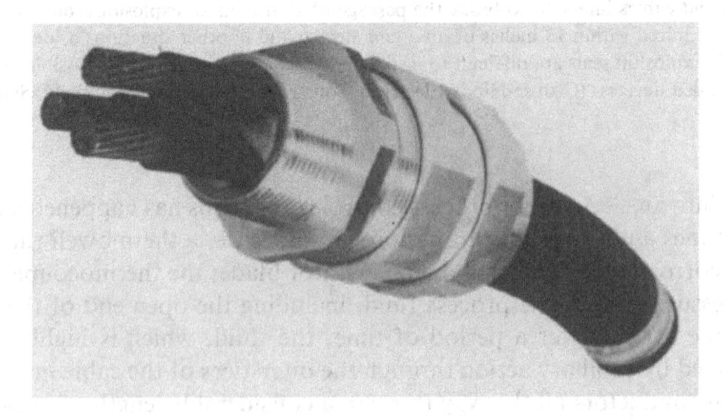

FIGURE 4–6. Cable seal. This is a type TMCX cable used for type metal clad (MC), cable. Inside the metal coupling is a seal formed by an epoxy, putty-like material, instead of the concrete-type sealing material typically used. The seal, therefore, is easier to install and can be disconnected. This new type of seal illustrates the improvements being made by the manufacturers of explosionproof equipment. (Crouse-Hinds Electrical Construction Materials Company, Syracuse, N.Y.)

process fluids from entering the electrical conduit or cable system of canned pumps and process connections that use a single seal, an additional approved barrier seal should be provided. The additional seal will prevent process fluid from entering the conduit system beyond the seal if the primary process seal fails. Vents or drains should be provided so the leakage is obvious. The double seal and drain system has been used in various forms at the thermocouples and canned pumps.

Another alternative is to provide an intermediate junction box so as to break the path from the process to the control room. The inside of the junction box should be considered Division 2. The disadvantage of the intermediate junction box approach is that it does not reveal that failure and leakage have occurred. Although Article 500 is concerned with flammable materials, the concern for propagation through process connections applies to any process fluid, especially toxic fluids. It seems incredible that an errant conduit could transmit flammable or toxic materials into a process control room or electrical equipment rooms or enclosures.

PURGING AND PRESSURIZATION

The *NEC* and international standards recognize the use of purging and pressurization as a method to reduce the classification within instrument and electrical enclosures, rooms, or buildings located in hazardous (classified) locations. Following the requirements in ISA Standard 12.4 and NFPA 496, it is possible to use equipment and installation practices of a lesser classification, that is, Division 2 instead of 1, unclassified instead of Division 2, and unclassified instead of Division 1.

Purging reduces the classification within an enclosure by sweeping away any flammable vapors or gases that may be present and applying a positive pressure to prevent the entry of any flammables that may be present on the outside of the enclosure (see Figure 4–7).

Purging and pressurization (PP) does not change the classification outside the enclosure. They can also be applied to Class II, combustible dust locations. The *NEC,* in Article 500, paragraph 506-2, indicates that classified locations can be eliminated or limited by adequate positive pressure ventilation from a source of clean air with effective safeguards against ventilation failure, and it refers to NFPA Standard 496-1986, *Purged and Pressurized Enclosures for Electrical Equipment in Hazardous Locations.*

Section 501-3 covering meters, instruments, and relays in Class I, Division 1 locations indicates that approved enclosures include explosionproof and purged and pressurized enclosures. Section 501-8 indicates motors and generators for Class I, Division 1 locations can be a totally enclosed type, supplied with positive pressure ventilation from a source of clean air with

FIGURE 4-7. Purging and pressurization instrumentation: various pressure instruments provided for the pressurization of enclosures to reduce the classification inside an enclosure. The minimum pressure should be 0.1 inch of water. These instruments are usually connected to an instrument air supply. Pressure regulators, as indicated, reduce the pressure, and pressure switches are provided to indicate that the pressure is too low. (Bebco Industries, Inc., La Marque, Tex.)

discharge to a safe area and so arranged to prevent energizing of the machine until ventilation is established and the enclosure has been purged with a minimum of 10 volumes of air and also designed to energize the equipment automatically if the air supply fails. They can also be the totally enclosed, inert gas-filled type supplied from a reliable source of inert gas for pressurizing the enclosure and have devices to ensure positive pressure in the enclosure and designed to deenergize the motor or generator automatically in the event the gas supply should fail.

Purging and pressurization was first recognized as a way of installing instrument enclosures in hazardous areas by the ISA, Standards and Practices (SP) Committee SP 12. In 1960 this committee issued ISA 12.4, *Standard Practice Instrument Purging for Reduction of Hazardous Area Classification,* as a tentative standard; in 1970, it produced the final version.

This standard is limited to instrument enclosures and introduced the concept of three levels of protection (X, Y, and Z) against the loss of pressurization commensurate with the degree of declassification. Type-Z purging is defined as the type of purging system that would permit a reduction of classification from Division 2 to unclassified. The protective requirements for this type of purging are as follows:

1. Equipment shall not be energized until at least four enclosure volumes of purge gas have passed through the enclosure while maintaining an internal positive pressure of at least 0.1 inch of water.
2. Under normal conditions, neither the temperature of the external surface of the enclosure nor that of the purge gas leaving the enclosure shall exceed 80% of the autoignition temperature (in degrees Celsius).
3. Failure of an enclosure shall be detected by an alarm or an indicator, but safety interlocks need not be provided.
4. If an alarm is used, it shall be visual and audible and located so it is readily seen or heard.
5. If an indicator is used, it shall be located for convenient viewing.
6. An easily visible warning nameplate shall be mounted on the enclosure, visible before the enclosure, containing the following or an equivalent statement:

   Enclosure shall not be opened unless the area is known to be nonhazardous or unless all devices within have been deenergized. Power shall not be restored after enclosure has been opened until enclosure has been purged for *X* minutes. (This time should be long enough to permit cooling to a safe temperature.)

7. The maximum surface temperature of internal components shall not exceed 80% of the autoignition temperature (in degrees Celsius) of the flammable vapor of gas involved.

Type-Y purging is defined as the type of purging system that would permit a reduction of classification from Division 1 to Division 2. The protective requirements for this type of purging include the Type-Z requirements and the following:

1. Internal components and wiring shall meet Division 2 requirements.
2. The enclosure thickness and material and size of the circuit protection must be such that any wiring fault will not burn through the enclosure nor allow the surface temperature to exceed 80% of the ignition temperature.

Type-X purging is defined as the type of purging system that would permit a reduction of classification within the enclosure from Division 1 to unclassified. This two-level reduction in classification therefore requires additional measures to ensure pressurization. They include the following:

1. Four volumes of purge gas must pass through the enclosure, with the internal pressure at least 0.1 inch of water, as determined by a timing device used to prevent energizing electrical equipment in the enclosure until the purge cycle is complete.
2. A cutoff switch, either flow or pressure activated, that will remove power automatically from circuits within the enclosure not suitable for Division 1 must be provided.
3. Motors, transformers, and similar equipment subject to overload shall be provided with overtemperature devices to deenergize the equipment on overload.

The latest and most complete standard for purging and pressurization is NFPA 496, *Standard for Purged and Pressurized Enclosures for Electrical Equipment* (1986). The standard includes chapters on instrument enclosures, control rooms, power equipment, and analyzer enclosures; it includes dust areas as well as areas for vapor and gases. It uses the X, Y, and Z classification and is similar to ISA SP 12.4, except that it covers a broader area.

There are several detailed requirements in this standard that should be reviewed before applying a purge system. All systems require a minimum pressure of 0.1 inch of water to exclude flammable vapors, gases, or dusts. This is the equivalent of the static pressure corresponding to a 15 miles/hour wind velocity. Two-tenths inch of water has been used to design pressurization systems for enclosures. Pressurization systems for control, electrical, or analyzer rooms or buildings must be limited to permit personnel to open outside doors. Four-tenths inch of water has been used as this limit.

The reliability of the purge supply, integrity of the enclosure, and protection against entry of outside process materials, alarms, interlocks, and warning nameplates are important issues addressed in this standard. All windows should be constructed of tempered shatterproof glass. Instrument air or inert gas can be used to purge and pressurize enclosures. The air intake to these systems must be located where it is not subject to vapor releases.

Chapters 8 and 9 of the NFPA 496 address situations in which the source of flammable gas or vapor may be internal to the enclosure or room. Purge requirements depend on the external electrical classification, approval rating of the internal electrical equipment, magnitude of the internal release, whether the release is limited by the piping system, integrity of the piping or tubing system, and so on. Containment of flammable gases or vapors is a greater issue in these enclosed systems.

ADVANTAGES OF PURGING AND PRESSURIZATION

Purging and pressurization has the advantage of not depending on the *NEC* groups A, B, C, or D. Therefore, it can be used in classified areas where explosionproof enclosures are not available. It also can be used where specialized equipment is not available as explosionproof enclosures or as intrinsically safe or nonincendive. Packaged equipment and control panels sometimes fall into this category. It could also be used in a situation where the group changes frequently, such as a pilot plant or research facility. Pressurization also provides protection against the entry of dirt, moisture, dust, and corrosive chemical vapors.

Purging and pressurization can also be applied to explosionproof enclosures as an additional protective measure and to protect against the entry of dust, dirt, moisture, and corrosive vapors.

The servicing and maintenance of pressurized enclosures is easier than that of explosionproof enclosures, and so on. It is necessary only to see that the enclosure is pressurized, that all covers are in place, and that the enclosure is not damaged and has no openings. Any conduits into the enclosure should be sealed to prevent excessive purge gas flow.

Monitoring the purge pressure is an issue whether it is by an indicator, alarm, or interlock. Type-X purging requires interlocks to remove power from equipment inside the enclosure if the purge fails (either flow or pressure actuated). These safety interlocks should be designed, installed, and tested as fail safe and should be hardwired (not be part of a PLC).

Wherever inert gases are used, it is essential to protect people against asphyxiation. Warning signs and training procedures can be considered. Protection against excessive pressure is required in the event that regulators or other piping system components fail.

Purging and pressurization systems are usually individually designed. This design should follow the standards previously cited. "Electrical Safety: Designing Purged Enclosures," an article in *Chemical Engineering* (May 13, 1974) by engineers at the 3M Company provides guidance in designing systems.

The term *sealing* has a number of meanings when applied to hazardous (classified) areas. In section 501-3 of the *NEC,* general-purpose enclosures are permitted in Division 2 locations if the current interrupting contacts are enclosed in a chamber hermetically sealed against the entrance of gases or vapors. In section 501-5 (Sealing and Drainage), seals are provided in conduit and cable systems to prevent the passage of gases and vapors, to prevent the passage of flames from one portion of the electrical installation to another, and for canned pumps, process, or service connections that depend on a seal to prevent flammable or combustible fluids from entering the electrical system. The three types of sealing are to

1. Protect against the exposure to an outside gas or vapor-hermetically sealed contact
2. Prevent the propagation of an explosion from one part of the electrical system by conduit and cable seals
3. Prevent the transfer of process fluids in the event a process seal fails

A hermetically sealed device is sealed against the entrance of an external atmosphere; the seal is made by fusion. Hermetic sealing also protects contacts against water and corrosion. Examples of the devices that are hermetically sealed include reed relay contacts, mercury tilt switches, or mercury plunger relays. These devices are especially useful in Class 1, Division 2, Group B locations, where explosionproof devices may not be available.

In general, all electrical enclosures are "sealed" because they restrict the entry of outside gases or vapors. Electrical enclosures are designed to prevent the entry of water, vapors, dirt, and liquids into the enclosure to protect the wiring and components inside the enclosure. All electrical enclosures are sealed to restrict or prevent the entry of outside air in order to protect the internal components and wiring from moisture, dust, and corrosive vapors and liquids. It is essential for the components and wiring to survive in chemical plant environments. If an enclosure cover were left lose or a gasket seal were removed or damaged, the wiring and internal components would be exposed to water, chemicals, and so on, and would fail prematurely.

Article 501-15 indicates there shall be no exposed live parts in Class I, Division 1 or 2 locations, for example exposed terminal strips. In *Electrical Instruments in Hazardous Locations,* Magison (1978) provides an excellent discussion of sealing. Tests and calculations indicate that any industrial enclosure, certainly NEMA 3 or NEMA 4, can provide many hours of protection against the entry of a flammable gas. Even explosionproof enclosures with standard flange gaps are effective in restricting the diffusion of outside air–vapor mixtures to the inside of an enclosure.

INTRINSIC SAFETY

Intrinsic safety (IS) is a design concept that applies to low-energy instrumentation systems—typically 24 volt direct current—that permits these systems to be installed in Division 1 locations without the special rules. Intrinsic safety apparatus does not have to be explosionproof or purged and pressurized, or have explosion seals. Wiring for IS systems does not have to follow Article 500 but only needs to meet the requirements for ordinary locations.

Intrinsic safety systems must, however, be designed, installed, and maintained correctly. They are designed with multiple safety factors and tested under the most adverse conditions so the probability of a properly designed, installed, and maintained IS system causing ignition is essentially zero.

The following rules apply:

1. IS equipment and wiring are approved for location. They must be listed for the particular class and group.
2. IS wiring must be separated from other wiring.
3. IS systems shall not be capable of releasing sufficient electrical or thermal energy. This means energy from an arc, spark, hot surface, or hot wire cannot cause ignition.
4. Ignition cannot occur under normal or abnormal conditions. This includes opening, shorting, or grounding of field wiring and multiple component failures.

The 1990 *NEC* provided a special section, Article 504, dedicated to the installation of intrinsically safe systems. In particular, the article indicates the following:

1. The installation of IS apparatus, associated apparatus, and other equipment shall be in accordance with the control drawings.
2. Intrinsically safe apparatuses shall be permitted in general-purpose enclosures.
3. Ordinary location wiring methods can be used in Division 1 locations.
4. Conduit and cable explosion seals are required only to prevent the ingress of process fluids. Process-connected instrumentation is again the concern.
5. Conductors of IS circuits shall not be placed in any raceway, cable tray, or cable with conductors of any nonintrinsically safe circuit unless:
   - Separated by a distance of at least 2 inches and secured
   - Separated by a grounded metal partition or approved insulating barrier
   - The IS or non-IS circuit conductors are metal-sheathed or metal-clad cables
6. Conductors of IS circuits within enclosures shall be separated by at

least 2 inches from conductors of other systems and shall be secured so they cannot contact another terminal. Separate wireways are the preferred way of achieving segregation. Different IS circuits should be in separate cables or separated by a grounded metal shield or insulation.

7. Grounding requirements shall follow Article 250. Note that RP 12.6 emphasizes the special grounding requirements for barrier systems.

ISA RP 12.6-1987 is an updated and upgraded version of this standard and provides a basis for many of the IS code additions.

The following terminology is used in IS; refer to RP 12.6 for complete definitions.

Associated Apparatus Apparatus in which the circuits are not necessarily intrinsically safe, but affect the energy in the intrinsically safe circuits and therefore are relied upon to maintain intrinsic safety.

Control Drawing A drawing or other document provided by the manufacturer of the intrinsically safe or associated apparatus that describes the allowed interconnections between the intrinsically safe and associated apparatus.

Intrinsic Safety Barrier An electrical circuit intended to limit the energy (voltage and current) available to the protected circuit in the hazardous (classified) locations, under certain fault conditions (Figures 4-8 and 4-9).

Intrinsically Safe Circuit A circuit in which any spark or thermal effect is incapable of causing ignition of a mixture of flammable or combustible material with air under defined test conditions.

Simple Apparatus A device that will neither generate nor store more than 1.2 volts, 0.1 amperes, 25 milliwatts, or 20 joules. Examples are switches, thermocouples, light-emitting diodes, and RTDs.

RP 12.6 describes the installation requirements for IS systems. It includes the separation and segregation requirements for IS and non-IS wiring. It also includes the special grounding requirements for barriers. That is, a quality signal grounding system is required to minimize stray currents that might cause undesirable differences in potential. RP 12.6 also has guidelines for application of the entity concept. The entity concept provides a method to determine acceptable matching of associated apparatuses (barriers) and IS connected apparatuses that have not been evaluated as a system. It also provides a method to determine whether the cable is acceptable for the IS installation. This is accomplished by determining the maximum allowable cable capacitance and inductance.

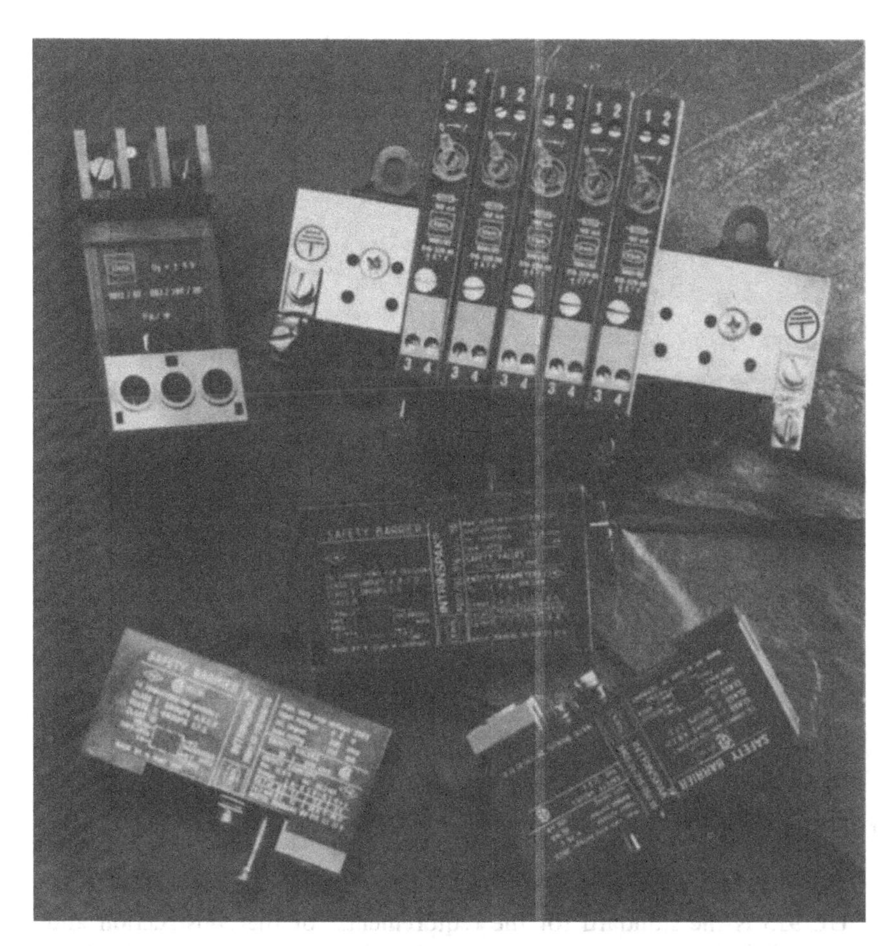

FIGURE 4-8 and 4-9. Intrinsic safety barriers. Barriers are electronic devices inserted into instrumentation wiring that enable field instrumentation to be part of an IS system. Various types of barriers are available from a manufacturer to match the instrument loop design. There are also significant differences in barrier design among various manufacturers, each taking different circuit design and packaging approaches. Note the barriers are labeled with information relating to the application of the particular barrier. (R. Stahl, Inc., Woburn, Mass.)

FIGURE 4-9.

UL 913 is the standard for the requirements for the construction and testing of electrical apparatus or parts of such apparatus that are intrinsically safe. This standard considers apparatus investigation for intrinsic safety, either as a system (field device in Class I, Division 1 location connected to associated apparatus) or separately using the entity evaluation. Entity evaluation provides a method to determine acceptable combinations of associated apparatuses and IS apparatuses by comparing electrical circuit parameters and other characteristics. The control drawing is the key document and is essential to the application of this method and the correct connection and grounding of apparatuses. It is required documentation in the 1990 *NEC*.

The fundamental requirement for IS circuits and apparatuses is that the energy available in the Division 1 location shall not be capable of igniting the flammable mixture, either by arcing, hot surfaces, or hot wires. For the purposes of evaluation, "normal" operation includes supply voltage at maximum rated value, any one of the field wires open, any two shorted, or any field wire grounded and other conditions. Evaluation of apparatuses can be by testing or calculation (comparison to tested ignition curves). This is done under fault conditions, which can be a single fault with an additional test factor of 1.5 applied to energy or two faults.

The evaluation considers all ignition sources including spark ignition from

- Breaking an inductive circuit
- Make-and-break resistive circuit
- Hot wire fusing
- Discharge capacitive circuit or thermal ignition from filaments, heating of a small gauge wire, or surface temperature of a component like a resistor

The comparison method uses the ignition curves for resistance, inductance, and capacitance circuits and for hydrogen (Group A and B), ethylene (Group C), and propane (Group D). Igniting currents for inductive and resistance circuits and ignition voltages for capacitive circuits are indicated for various electrode materials. The circuits must be capable of being analyzed for single-fault or two-fault conditions.

The spark ignition test method uses a spark test designed to promote spark ignition and achieve the maximum transfer of spark energy to the flammable mixture. The sensitivity of the test apparatus must be verified before testing. Ignition must not occur after hundreds of operations for apparatuses to pass the test.

The concept of intrinsic safety is based on the fact that for a given flammable mixture there is a minimum energy that must be injected into a flammable mixture in order for a flame to grow indefinitely and flame propagation to occur. For a given material and test condition there is a concentration for which the energy is minimum. The ignition energy versus volumetric concentration curve is U shaped. Intrinsically safe systems limit the energy release to a value below this minimum ignition energy even under the most adverse circumstances.

Sparking between energized parts or an energized part and ground does not provide total transfer of the spark energy to the flammable mixture. Some energy is always dissipated by the arcing contacts (or electrodes). In the spark testing described in various standards, the goal is to achieve effective transfer of the spark energy and therefore the minimum ignition energy.

The barrier method is the most popular and convenient method of applying intrinsic safety. It allows various types of equipment in the unclassified location to be coupled to equipment in the Division 1 location. Various types of equipment can be provided on the load side if the circuit characteristics are known. Barriers are designed so that if the circuits and equipment that energize the barrier on the nonhazardous side of the barrier fail and produce high voltage (up to 250 vac), the barrier will limit the energy in the hazardous side to a safe value.

On the hazardous side of the barrier, it is essential to verify that the stored energy in the load cannot cause ignition. If the inductance is too high, an ignition-capable spark could occur if the circuit is opened. If the capacitance is too high, an ignition-capable spark could occur if the wiring is shorted. The inductance and capacitance of the cable should be included in these calculations.

The application of the barriers requires consideration of the field wiring and the type of loop. Many types of barriers are available, depending on the grounding, loop current and voltage rating, polarity, and so on. Barrier resistance is also a factor that should be considered. Close coordination with the barrier vendor is essential.

Intrinsically safe barriers are available for most of the types of field devices. Intrinsically safe portable equipment for instance, combustible gas detectors and handie talkie radios, is also available. Intrinsically safe systems have the advantage that they can be maintained in a Division 1 location without deenergizing the equipment. In highly corrosive locations, plastic enclosures can be used instead of metal.

Nonincendive circuits and components can be used in control systems in Division 2 locations. Nonincendive circuits or components are not an ignition source under normal operation. A device without contacts is obviously nonincendive; however, the real value of the nonincendive concept is the recognition that instrument circuits can operate at such a low-energy level that at normal operation, they would have a very low probability of causing ignition. For instance, if an instrument alarm system switches 2 milliamps at 24 volts with a series 12 thousand ohm resistor, even under adverse circumstances, it could not cause ignition of many flammables.

Nonincendive equipment is equipment that in normal operating condition would not ignite a specific hazardous atmosphere in its most easily ignited concentration. The circuits may include make-and-break contacts releasing insufficient energy to cause ignition. Circuits not containing sliding or make-and-break contacts may operate at energy levels potentially capable of causing ignition.

The minimum ignition energy is a characteristic of a specific gaseous mixture. It is the smallest amount of energy that can cause ignition of the mixture at normal room temperature and pressure. Tests made at high voltage by discharging capacitors through pointed electrodes give the lowest energy measurements; such data are believed to approach the true minimum ignition energy. The ignition energy will decrease as temperature or pressure or both are increased.

Nonincendive circuits may include open, sliding, or make-and-break contacts releasing insufficient energy to cause ignition. Sliding or make-and-break contacts otherwise protected, such as by suitable sealing, purging, or oil immersion, are acceptable regardless of the power level.

Nonincendive equipment and wiring may contain circuits having energy levels potentially capable of causing ignition. Tube sockets, disconnect plugs, calibration adjustments, and so on, which are not disconnected in normal operation but are operated only for corrective maintenance, can not be considered make-and-break contacts and are acceptable regardless of power level if they are located within an enclosure.

Approval may be based on examination or ignition testing. Voltage and current in sliding or make-and-break contacts for given circuit constants may exceed the levels given if the circuit is shown to be safe by ignition testing.

Electrical equipment is nonincendive if, in its normal operating condition, voltage and current in sliding or make-and-break contacts for given circuit constants do not exceed 25% of the value of current determined from the appropriate figure or 25% of the value of the voltage as determined from the appropriate figure. Normal conditions include maximum supply voltage and extreme environmental conditions within the stated ratings of the equipment.

Equipment and wiring having an exposed surface at a temperature exceeding 80% of the ignition temperature in degrees Celsius of the specific hazardous material shall not be classed as nonincendive. For a very small surface, such as a straight fine wire, this temperature may be exceeded, provided it is shown that under normal conditions it will not ignite a specific hazardous atmosphere.

Based on the work of ISA SP 12, nonincendive circuits were added to the *NEC*. The 1987 *NEC* indicates that enclosures that contain make-and-break contacts of push buttons, relays, and so on, would not have to be in "approved" enclosures in Division 2 locations if the current-interrupting contacts are enclosed with a chamber hermetically sealed, immersed in oil, or nonincendive. This applies to meters, instruments, and relays (Article 501-3) and to signal, alarm, remote-control, and communication systems (Article 501-4).

The concept of nonincendive was further developed and expanded in ISA standard, *Electrical Equipment for Use in Class I, Division 2 Hazardous (Classified) Locations,* ISA SP12.12 (1984). This standard includes definitions for *nonincendive circuit, field wiring,* and *component.*

Nonincendive circuit is a circuit in which any arc or thermal effect produced under normal operating conditions of the equipment is not capable of igniting the specified flammable gas or vapor-in-air mixture. *Nonincendive circuit field wiring* is wiring that, under normal operating conditions of the equipment, is not capable of igniting the specified flammable gas or vapor-in-air mixture by opening, shorting, or grounding. *Nonincendive component* has contacts for making or breaking an incendive circuit, where either the contacting mechanism or the enclosure in which the contacts are housed is so constructed that the component is not capable of propagating ignition.

SP 12.12 indicates that any contact device (for instance, a relay, circuit breaker, potentiometer, or switch) can be in a nonincendive circuit, a nonincendive component, or a sealed device.

The standard provides guidance concerning marking, surface temperature requirements, and field wiring. Evaluation of nonincendive circuits can be based on testing similar to the test apparatus and techniques used for intrinsic safety or by comparing calculated or measured values of voltage, current, and associated inductances and capacitances to the same ignition curves used for IS. If the values for current and voltage are less than the curve for the same circuit and flammable, the circuit is nonincendive. The process is similar to IS evaluation, except component fault conditions are not considered, greatly simplifying the process.

There are still significant safety factors. Ignition testing occurs at the most easily ignitible concentration, the test apparatus and procedure are designed to maximize the possibility of ignition, and field wiring failures are included in the evaluation.

Nonincendive component is a newer concept. The component is to be tested in an explosion chamber with the appropriate gas at its most easily ignitible concentration and operated 50 times at 150% of the rated load; for the component to be nonincendive, no ignition should occur during this testing.

OIL IMMERSION

Oil immersion of electrical contacts is recognized as an acceptable protective technique for Division 2 locations in Article 501 of the *NEC* in the following sections:

501-3 Meters, Instruments, and Relays
501-5 Sealing and Drainage
501-6(b) Switches, Circuit Breakers, Motor Controllers, and Fuses

Protection against ignition is dependent on the immersion of contacts in oil and maintaining an adequate oil level in the enclosures. Oil immersion has been applied to motor push buttons, lighting power panels, and other power equipment. It is highly dependent on proper installation, maintenance, and inspection. Underwriters Laboratories Standard 698 provides the requirements for oil-immersed apparatus.

CONCLUSION

Electrical facilities in Class I locations must, as a minimum, be provided with the protective measures described in Article 500 of the *NEC*. The enclosures and installations must be the appropriate class, group, and T number for the particular flammable involved. Compliance with Article 500 is not easy to accomplish. It involves not only the type of enclosure, but also the installation, including wiring method, grounding, and sealing. The protective measures include explosionproof enclosures, intrinsically safe systems, purged and pressurized enclosures, nonincendive circuits, hermetically sealed contacts, oil immersion of contacts, and other installation and wiring practices. The requirements in Division 1 locations are especially onerous.

The choice of various options depends on a number of factors, including type of equipment, environment, and maintenance practices, and these should all be considered before choosing a type. Wherever possible, select listed and labeled equipment. In Division 1 locations, this is generally a requirement; in Division 2, it may or may not be.

Always ensure that the correct T number is specified. In Division 2 locations where explosionproof enclosures are not always required, it is essential to verify that all surface temperatures (internal or external) under normal conditions do not exceed 80% of the autoignition temperature of the particular gas or vapor involved. This applies to motors, lights, solenoids, heaters, and other heat-generating apparatus.

Explosionproof apparatus has a proven record of safety, but it must be maintained correctly. The integrity of the flame path at the cover is especially important. If the flame path is scratched by a screw driver or a gritty material, the explosionproof integrity may be compromised. There are trade-offs between the explosionproof design and watertightness, but some of the newer enclosure designs using an O ring seal are a significant improvement in this respect.

Intrinsic safety has a high reliability but must be wired with the separation requirements indicated in Article 504. Intrinsically safe systems must be grounded correctly, and the barrier circuit parameters must be compatible with the instrument loop. Close coordination with the barrier manufacturer is essential. Intrinsically safe systems also have the advantage that they can be maintained while energized in a classified location.

Purging and pressurization has a number of advantages: It does not depend on group classification; it keeps moisture, dust, and chemical vapors outside of the enclosure; it can be applied to equipment not listed for a group; and it can be maintained in a classified location if the purge rate is sufficient to keep flammable vapors from entering the enclosure. Purge systems must be designed for a particular application. The alarm systems for loss of pressure must be checked periodically.

Nonincendive circuits can be used in Division 2 locations and are easier to use than IS systems.

There are special cabling and grounding requirements for classified locations, but the most difficult requirement is the installation of explosion seals. Explosion seals require special care: Only the materials provided by the manufacturer can be used, the conductors must be separated, and the seals must be poured when the temperature is above a certain limit. Whenever possible, factory-sealed devices should be used to avoid sealing.

REFERENCES

Angehrn, J., W. Short, and W.O.E. Korver. 1975. Design and Construction Guidelines for Electrical Systems in Hazardous Locations. *Electrical Construction and Maintenance* (Nov.):75–82.

Bossert, J., and R. Hurst. 1986. *Hazardous Locations: A Guide for the Design, Construction and Installation of Electrical Equipment.* Rexdale, Ontario: Canadian Standards Association.

Crouse Hinds Electrical Construction Materials. *1990 Code Digest.* Syracuse, N.Y.

Magison, C. 1978. *Electrical Instruments in Hazardous Locations.* Instrument Society of America, 3d ed. rev.

5

Dust Electrical Safety in Chemical-Processing Facilities

In less than one week in December 1977, two grain elevators in modern grain export terminals exploded, killing 54 people and causing estimated direct losses of more than $50 million. In a recent situation, an operator died after inserting a trouble light inside a plastic feed tank, causing a plastic dust cloud ignition. Such dust explosions and fires occur with regular frequency in the grain and other industries as well as in chemical-processing facilities.

DUST EXPLOSIONS AND FIRES

Dusts are deceptive, and their familiarity sometimes masks their potential for fire and explosion. Sugar, flour, and coal dust are familiar and seemingly nondangerous materials, but when they are processed on an industrial scale, the potential for fire and explosion is ever present.

A dust explosion can begin as a small fire or as dust melting. This can cause further heating and turbulence of the air, which disperses more dust, which in turn provides fuel for more heating, burning, and air pressure buildup. The initial burning and pressure rise can dislodge dust accumulations on building structures, piping, ductwork, and machinery, which provide more fuel and can result in secondary and subsequent explosions. Witnesses of dust fires have reported sensing air-pressure changes and seeing burning dust clouds. The final explosion is a rapid burning of a cloud of dispersed dust and a release of thermal energy in a rapidly rising air-pressure wave.

Dust layers can produce dust clouds if the layers are sufficiently dispersed or agitated. Dust layers can also melt, char, and burn and propagate a flame, either by excessive temperature if the dust layer has settled on a hot surface, such as an electrical motor or lighting fixture, or by sparking material from a welder or a fault in electrical equipment falling on the dust layer.

Dusts are a necessary part of many chemical-process facilities. They are usually not the final product but occur as fines in milling, cutting, and other mechanical operations. Almost any chemical dust can be made to explode if the dust is fine enough, the dust cloud dense enough, or the dust layer thick enough and a strong ignition source—either a spark or hot surface—is present.

ELECTRICAL CAUSES OF DUST IGNITION

The possible ignition sources for dust clouds and layers include sparks produced by welding tools and equipment, static electricity, lightning strikes, and electrical equipment and wiring. Articles 500 and 502 of the *NEC* have the avowed purpose of preventing electrical equipment and wiring from igniting combustible dusts. This discussion is concerned with such preventions.

As with Class I flammable vapor and gas locations, it is necessary to classify a dust location as Class II, Division 1, 2, or unclassified and the correct electrical enclosure as Group E, F, or G. After a Class II location is classified, the protective measures for electrical equipment and wiring are described in Article 502 of the *NEC.*

CHARACTERISTICS OF DUST EXPLOSION

As with Class I vapors and gases, it is first necessary to determine if the material is combustible and, if so, to what degree. This can be done by testing the individual dust or by referring to test data on similar dusts and assuming your dust will act in a similar way. Testing usually follows the practices established by the U.S. Department of Interior, Bureau of Mines. In the early sixties, the Bureau of Mines issued a number of reports on the explosibility of dusts in various industries. RI 5624 describes the laboratory equipment and test procedures for evaluating explosibility of dusts. The dust explosion parameters considered are

Minimum dust cloud explosive concentration
Ignition temperature of a dust cloud
Ignition temperature of a dust layer
Electrical energy for ignition of a dust cloud
Electrical energy for ignition of a dust layer
Relative flammability of a dust cloud
Limiting oxygen concentration
Pressure and rate of pressure rise

The Bureau of Mines has published reports of investigations of the explosibility of metal powders, carbonaceous dusts, and agricultural, plastic, chemical, drug, dye, pesticide, and other miscellaneous dusts. Of particular interest to chemical processing is RI 7132, *Dust Explosibility of Chemicals, Drugs, Dyes, and Pesticides* and RI 5971, *Explosibility of Dusts Used in the Plastics Industry.*

Two indexes are defined that provide a measure of relative explosibility compared to Pittsburgh coal dust. One is the *ignition sensitivity,* which equals the ratio of the *dust cloud ignition temperature* times the *dust cloud minimum ignition energy* times the *minimum dust cloud explosion concentration* for Pittsburgh coal dust to the sample dust. The other is the *explosion severity,* which equals the ratio of the *maximum explosive pressure* times the *maximum rate of explosive pressure rise* for the sample dust to Pittsburgh coal dust.

The overall Index of Explosibility is the product of the *explosion severity* and the *ignition sensitivity.* If their ratios are one, it implies the dust sample is equivalent to Pittsburgh coal dust as far as explosibility is concerned. Explosion hazards are further classified as weak, moderate, strong, and severe according to this ratio as established by the Bureau of Mines.

The report on explosibility of plastics dust reveals that composition and chemical structure are important factors and that there is variation with the dust samples even if the dust appears to be the same chemical composition.

Particle shape is also important. Irregularly shaped particles present a greater explosion hazard than do spherical particles. For instance, spherical particles of one plastic had an explosibility index of less than 0.1, indicating a weak explosion hazard, whereas irregularly shaped particles of the same material had an explosibility index as greater than 10, indicating a severe hazard.

Calculations of explosion severity were based on a dust cloud concentration of 0.5 ounce per 1 cubic foot. Other dust cloud concentrations affect explosibility as shown in the following test data for one dust:

Dust Cloud Concentration (oz./cu. ft)	Pressure Rise (PSIG)	Maximum Rate of Pressure Rise (PSI/sec)
0.10	30	450
0.20	66	2800
0.50	89	4100
1.00	103	3600
2.00	113	2900

Bureau of Mines reports indicate that the basic chemical structure governs explosibility, and incorporation of halogens, chlorides, and fluorides works in the direction of decreased hazard.

The tests were based on dusts that pass through a 200 sieve (74 microns particle size). Dust particle size is a major factor in dust explosions. The smaller the dust particle size, the higher the probability of dust explosion. NFPA publication 69 defines a combustible dust as finely divided combustible material with a particle size 420 microns or less (material passing a No. 40 sieve) in diameter that can produce fire and explosion when dispersed or ignited. Coal dusts as large as 840 microns, however, have been involved in explosions.

In chemical-process operations there is a distribution of particle sizes, and although some of the particles may be pellet size and too large to explode by a reasonable ignition source, processes usually produce fines that may be explodable.

The results of the Bureau of Mines testing is dependent on the test equipment and methodology. For instance, the pressure and rate of pressure rise depend on the test vessel used. This is discussed in *Explosions* by W. Bartknecht. The dust layer thickness used to determine the ignition temperature of a dust layer is one-half inch. Ignition temperature of a dust layer decreases with layer depth and time of exposure. For many industrial situations this may be too low; for some it may be too high.

Measurement of ignition energies uses capacitive discharge circuits that achieve maximum transfer of energy to the spark. How relevant this circuit is to actual sparking in an electrical circuit remains to be seen.

NATIONAL ELECTRICAL CODE AND COMBUSTIBLE DUSTS

The *NEC* in Articles 500 through 503 covers the requirements for electrical wiring systems and equipment at all voltages where explosion and fire hazards may exist due to

Flammable gases or vapors Class I
Combustible dusts Class II
or
Ignitable fibers or flyings Class III

Dust locations are entirely different from flammable gases or vapors locations. The classification criteria, enclosure groups, and protective techniques are all different. The very nature of the process is different:

- Dusts accumulate; vapors and gases dissipate.
- Dust clouds and layers can be easily seen.
- Dusts can settle on motors and lighting fixtures and the blanketing effect can cause layer ignition.

Hot surface ignition is much more likely with dusts than with vapors; on the other hand, spark ignition of a dust cloud is less likely than for a vapor. Dust spark ignition energies are higher than that of vapors.

The grouping of explosion characteristics is based on dust tightness of shaft openings and joints of assembly to prevent entrance of dust in the dust-ignitionproof enclosure, the effect of blanketing of layers of dust on equipment that may cause overheating, ignition temperature of the dust, and conductivity of the dust. As with vapors and gases, it is necessary that equipment be approved for the particular group. Historically, there were three groups of dusts classified according to the type of dust:

- Group E included metal dusts like aluminum and magnesium.
- Group F included carbonaceous dusts, with carbon black, charcoal, coal, or coke dusts.
- Group G included flour, starch, or grain dust. Plastic and chemical dusts were usually in Group G.

These definitions were added to the 1978 *NEC*. In the 1981 *NEC*, dust conductivity criteria were added so the 1987 *NEC* defines Groups E, F, and G as follows:

Group E Metal dusts regardless of resistivity or other combustible dusts having similarly hazardous characteristics having a resistivity of less than 10^2 ohm-centimeter

Group F Carbonaceous dusts having a resistivity greater than 10^2 ohm-centimeter but equal to or less than 10^5 ohm-centimeter

Group G Combustible dusts having resistivity of 10^8 ohm-centimeter or greater

Class II locations are divided into three risk classes: Division 1, Division 2, and unclassified. Division 1 locations include those where dust is in the air under normal operating conditions in quantities large enough to produce ignitable or explosive mixtures, or where mechanical failure or misoperation could produce ignitable mixtures, or where mechanical failure or misoperation could produce ignitable mixtures and at the same time an electrical failure that could result in a source of ignition, or where electrically conductive dusts may be present in hazardous quantities.

Division 2 locations include areas where combustible dust is not normally in the air and dust accumulations are not large enough to affect the normal operation of electrical equipment. As with Class I (vapor and gas locations), the definitions of Division 1 and 2 are in broad terms. The criteria for classification are defined in NFPA 497B and ISA S12.10.

What is also necessary and considerably more important than vapor and gas locations is that the surface temperature of lighting fixtures, motors, and other heat-generating electrical equipment not be higher than the layer ignition temperature of the particular dust layer. T numbers are used as indicated in *NEC* Table 500-3(b). The T number should always be less than the layer ignition temperature of the particular dust. For instance, if a plastic dust melts, chars, and ignites at 90°C, a T6 number (maximum temperature 85°C) should be used. Note that this temperature is lower than the temperature formerly acceptable for Group G, 120°C.

NFPA 497M-1986, *Classification of Gases, Vapors, and Dusts for Electrical Equipment in Hazardous (Classified) Locations* lists cloud or layer ignition temperatures for many dusts. For most dusts, the layer ignition temperature is lower than the cloud ignition temperature, but for a few dusts cloud ignition temperature is lower.

NFPA 497M-1986 includes dusts whose ignition sensitivity is equal to or greater than 0.2 and explosion severity is equal to or greater than 0.5. The dust data are based on the Bureau of Mines reports, but due to variations in dust explosibility—even of dusts of the same chemical composition, it is far better to test dust samples from the actual process operation.

The requirements for electrical installations in Class II, combustible dust, locations are in Article 502 of the *NEC.* The principles and requirements are quite different than those of Class I, flammable vapor and gas locations. Dust-ignitionproof and dust-tight enclosures are required instead of explosionproof enclosures. Explosionproof enclosures are not acceptable in dust areas. In some situations, the enclosure can be made dust ignitionproof, and the enclosure may be listed for Class I and II locations. Dust-ignitionproof enclosures prevent dusts from entering the enclosures under adverse testing conditions and do not allow heat-generated arcs or sparks inside the enclosure to ignite dust on the outside. Dust-ignitionproof enclosures are not intended or tested to contain an explosion; they are designed and tested to keep combustible dust on the outside.

In class II locations, heat-generating equipment like motors, lighting fixtures, and heaters must not develop surface temperatures that could cause dehydration or carbonization of a dust even if they are blanketed.

The UL testing of dust-ignitionproof equipment includes placing the equipment in a test chamber and circulating dust. Penetration of dust inside the equipment is checked, and the equipment is heat cycled to ensure that the surface temperature does not char any external dust accumulations on the equipment. Dust must not enter the enclosure during these tests. The surface temperature limits are 200°C for Group E, 200°C for Group F, and 165°C for Group G.

Tests for motors include the motor under load and locked rotor and single phasing. Heating of motor bearings is also considered. The types of dust used for testing is grain dust for Groups F and G and magnesium dust for Group E.

Intrinsically safe systems can be used in Class II locations. The same UL standard is used as for Class I locations. Spark testing, however, is not required if the system meets the requirements of Class I locations. This is acceptable because of the higher ignition energies required to ignite a dust cloud and the difficulty in performing IS spark testing for dusts.

The nonincendive concept is also permitted in Class II, Division 2 locations. Spark testing is not required if the circuit is safe for a Class I location. IS and nonincendive systems have surface temperature limitations based on the particular group. Pressurization, per NFPA 496, can also be used in Class II locations, with minor variations over Class I requirements.

Hermetically sealed contacts and oil immersion of contacts are permitted in Division 2 locations. Dust-tight enclosures, however, are the most convenient method of protecting Division 2 locations and also provide environmental protection against water and rain. *Dust tight* means constructed so dust will not enter the enclosure. The code refers to NEMA Standard 250-1979, *Enclosures for Electrical Equipment,* Paragraph 250.5.05 as to type of dust testing. This standard indicates that NEMA Types 3 and 12 enclosures are designed to pass this test and therefore may be acceptable in Division 2 locations.

In Division 1 locations the following rules are indicated in Article 502:

1. Transformers must be located in vaults. Transformers are not permitted in metallic dust areas.
2. Conduit on MI cable are required. Cable trays are permitted as support for MI cable.
3. Sealing is required but not Class I explosion sealing. The purpose of

dust sealing is to prevent transmission of dusts into a dust-ignition-proof enclosure.

4. Motor controllers are required to be dust ignitionproof.
5. Motors are required to be dust ignitionproof or pipe ventilated. There are a number of requirements for ventilating piping.
6. Dust-tight enclosures are permitted for control equipment if the contacts are sealed.

In Division 2 locations, in general, dust-tight enclosures are required, totally enclosed fan-cooled motors are permitted, and cable trays can be used if the cables are in only one layer and spaced one cable diameter apart. Special precautions are necessary when metallic dusts are involved. The grounding requirements are similar to Class I location requirements.

CLASSIFICATION OF DUST LOCATIONS

The definitions of Class II, Division 1 and 2 locations in Article 500-6 are in general terms and therefore difficult to apply to individual situations. Unlike Class I locations, there are no area classification figures in any of the NFPA standards for combustible dusts. Until recently, the only standard for dust classification was ISA Standard S12.10, *Area Classification in Hazardous Dust Locations* (1988). This standard was developed in the early 1970s and was updated in 1988. It presents a measurement approach to dust classification.

The standard introduced the concept of minimum hazardous dust layer thickness and measurement techniques for dust cloud density, dust layer thickness, and dust resistivity. In its revised form, it is the best resource for dust classification.

The minimum hazardous dust layer thickness is the smallest thickness that will not propagate a fire or flame. It is measured by applying a heated wire to layers of dust to determine the smallest layer thickness that will not propagate a flame. This thickness is relevant to a situation where sparking material from welding or a fault in electrical equipment could fall on a dust layer and possibly cause the layer to continue burning. Whether or not the burning continues depends on the type of dust and the layer thickness. For instance, a cold layer of grain elevator dust 44 mills thick will not continue to burn if the ignition source is removed, whereas a layer 66 mills thick will continue to burn. The standard also describes a method to measure dust cloud weight concentration and dust layer thickness.

The standard considers Class II, Division1 locations to be any of the following:

1. Locations where a dust cloud is present under normal conditions at an explodable concentration.
2. Significant dust accumulations are present for extended periods of time due to inadequate housekeeping or dust removal systems. Extended period of time is considered to be 0.1% of the time. Dust accumulators are considered significant if the dust layer thickness exceeds one-eighth inch.
3. The location is Division 2, and the dust is conductive (low resistivity). An electrically conductive dust can produce heating and possibly ignition of the dust if allowed to bridge between energized electrical terminals or an energized terminal and a grounded enclosure.
4. Mechanical failure or unusual operation of process equipment might simultaneously result in ignitable mixtures and cause an electrical failure that could cause ignition.

Division 2 locations are considered to be any of the following:

1. Locations where the frequency of occurrence of a combustible dust cloud is low. Low is considered to be more than 0.01% of the time but less than or equal to 0.1% of the time.
2. Locations where the dust accumulations exceed the combustible dust layer thickness or are great enough to obscure the color of the floor but are not large enough (greater than one-eighth inch) to justify a Division 1 classification.
3. Situations where dust is processed in an enclosed system but an occasional leak source occurs at flexible connections, rotary feeders, or other leak sources.

The standard also provides a method of measuring dust resistivity using the test apparatus indicated in Figure A-11 in S 12.10.

Test data for both resistivity and dielectric breakdown are presented for several types of dusts. The results are summarized in figure A-18 in S 12.10. The results indicate that dusts fall into three categories with respect to resistivity: high-resistivity insulator dusts like cornstarch and flour; medium-resistivity carbon dusts; and low-resistivity metallic dusts. Resistivity is a nonissue for most chemical and plastic dusts. The most critical issue for these dusts is layer ignition temperature.

The NFPA has issued a recommended practice for classifying dust. This standard, NFPA 497B, provides classification figures and should be used as an additional resource to ISA S 12.10.

CONCLUSION

Electrical facilities in Class II, combustible dust, locations must not cause dust explosions or fires. The dust explosion problem is an entirely different problem than a vapor or gas explosion problem. The basic phenomenon is different, as is the method of classification and the protective measures. Dusts accumulate, whereas vapors dissipate.

Dust classified areas can be reduced or eliminated by containment of the dust and good housekeeping. Dust classification can be done by inspection using the one-eighth-inch criteria. If the dust layer is greater than one-eighth inch, the location should be Division 1; if the layer thickness is less than or equal to one-eighth inch but still visible to the degree that the color of the floor is obscure, the classification is Division 2.

Dust accumulations should be viewed as fuel waiting to be ignited. Dust layer ignition by dust cloud dispersion, layer heating, or sparking material falling into a dust layer are the critical issues. Test data of dust layer thickness versus time to ignite are useful in defining the safe surface temperature for motors, lights, heaters, and so on, in dust classified areas.

Chemical dusts are Group G, and the maximum surface temperature permitted by the *NEC* is 120°C. Tests, however, indicate that some plastic dusts ignite in layers at lower temperatures; therefore, it is essential to test and define the safe surface temperature for a given dust. Dust conductivity is a nonissue in chemical dust facilities. It is expected that the conductivity numbers will be removed from Article 500 and that only metallic dusts could be considered conductive.

REFERENCES

American National Standards Institute/Instrument Society of America. 1988. ISA Standard S 12.10. *Area Classification in Hazardous (Classified) Dust Locations.*

Baggs, G. Safe Electrical Design Practices For Coal Handling Facilities. IEEE Paper PCI-82-72. 1982.

Bartknecht, W. *Explosions* Berlin/Heidelberg/New York: Springer-Verlag: 1981.

6

Electrical Safety in Chemical Processes

Electrical safety is important in any location or situation, but it is especially important in chemical processes because of the hazardous nature of the materials and the severe environmental and operating conditions.

Chemical processes use flammable gases and vapors and combustible dusts, which can be ignited by electrical equipment and wiring (as discussed in Chapter 5), as well as by lightning discharge and static electricity sparks. Lightning discharges can damage structures and buildings and cause momentary power blips or sustained damage to electrical power systems, resulting in major power outages. Both lightning discharges and static electricity can disturb and damage process control and instrumentation systems.

Chemical processes present the most severe operating and environmental conditions for electrical systems. Corrosive and toxic materials in chemical plants attack electrical conduits, panels, connections, cables, and people. Many plants are outdoors and subject to rain, dust, dirt, wash down, and wide temperature extremes. Plants on the Gulf Coast are subject to salt spray and condensation. In those plants, condensation inside conduits, boxes, instruments, power panels, motors, and so on, is a difficult problem. Grounding, protecting circuit and equipment, selecting electrical enclosures, and wiring systems involve technically complex issues and require a high level of electrical engineering capability. The continuity and reliability of electrical power to process equipment motors, heaters, lighting, and other electrical equipment are key issues in chemical processes.

Power interruptions can range from a momentary power blip or dips caused by lightning, power system switching, power system faults, and large motor starting to major long-term outage due to equipment or other power system failures. The result of a loss of power can range from a disruption of production to an explosion, a major release, or equipment failure. It depends on the type of process and equipment and the duration of the power outage.

Another factor that adds to the severe operating conditions in chemical processes is the fact that many chemical plants operate 24 hours a day, 7 days a week. This situation makes maintaining and repairing equipment difficult. In some cases, electrical equipment and wiring must be worked on while they are energized. Special precautions and work practices are required in these situations.

Operators, for example, may be exposed to special electrical hazards in locking and tagging out electrical equipment. Therefore, specialized electrical equipment and practices are required as a part of vessel entry. In addition, electrical equipment and wiring must be designed so that if a wiring or insulation failure occurs, the sparking and arcing fault current is of short duration and is contained within the electrical equipment. Electrical systems that have the correct circuit and equipment protection (fusing and breaker setting) and are properly grounded do not allow failures to result in burn downs of facilities.

To address these and other electrical safety issues, the following topics will be discussed in this chapter:

Electrocution and personnel safety
Lightning protection for chemical-process facilities
Static electricity as an ignition source
Protection of electrical systems
Electrical power reliability and quality
Cable systems for chemical-process facilities

ELECTROCUTION AND PERSONNEL SAFETY

Human beings are extremely sensitive to very small values of electrical currents that may pass through parts of the body as a result of accidental contact with electrically energized conductors. This is because the body is a conductor and functions by millivolt electrical potentials developed within different parts of the body and transmitted within the nervous system. Muscular action, heart action, breathing, and virtually every body action depend on an elaborate and sensitive electrical system that can be upset or destroyed when an external electrical current is superimposed on it.

Tests on people and animals have been performed that provide quantitative data on the sensitivity of the human body to electrical current. As shown in Table 6-1, the smallest perceptible current is about 1 milliamp or one-thousandth of an ampere. The given currents are typical values based on a number of tests or, in some cases, on tests on animals and the results

TABLE 6-1. Effects of Electrical Current in the Human Body

Current	Reaction
1 Milliampere	Perception level. Just a faint tingle.
5 Milliamperes	Slight shock felt; not painful but disturbing. Average individual can let go. However, strong involuntary reactions to shocks in this range *can* lead to injuries.
6-25 Milliamperes (women) 9-30 Milliamperes (men)	Painful shock, muscular control is lost. This is called the freezing current or "let-go"* range.
50-150 Milliamperes	Extreme pain, respiratory arrest, severe muscular contractions.* Individual cannot let go. Death is possible.
1,000-4,300 Milliamperes	Ventricular fibrillation. (The rhythmic pumping action of the heart ceases.) Muscular contraction and nerve damage occur. Death is most likely.
10,000– Milliamperes	Cardiac arrest, severe burns and probable death.

*If the extensor muscles are excited by the shock, the person may be thrown away from the circuit.

extrapolated to human beings. If the current is alternating (AC), the feeling is a tingling sensation; if it is direct current (dc) the feeling is a slight warmth. The actual effect will vary for each person, depending on the person's sex, age, and weight and the condition, duration, and frequency of the current. This is true for all the current values given.

Muscular action can be influenced by an external electrical current. With respect to this topic, there are two types of muscles: flexor muscles, used for gripping, and extensor muscles, used for hitting or kicking. If extensor muscles are involved in the path of the external electrical current, the victim might be knocked off a ladder or into a hot object or energized wiring. This secondary, involuntary movement could result in a serious injury. This current would be in the range of 5 milliamps ac. Higher currents above the let-go would not allow a person to release a grip if he or she were holding on to an energized object. In essence, the person is frozen to the object.

In one test, let-go currents were measured for a number of individuals by passing a test current from a handheld electrode to a second electrode strapped to the arm. For obvious reasons, the heart path was not involved. The tested range of values for men was 9.7–21.6 milliamps ac; the tests for women were somewhat lower, about 6 milliamps ac. This research established the basis for the operating current of Ground Fault Circuit Interruption (GFCI). This device, called a people protector, is required by the *NEC* in wet locations.

If the current passes through the chest region and exceeds 50 milliamps, the chest muscles contract so breathing is stopped. This condition is referred to as respiratory paralysis. If the current persists, collapse, unconsciousness, and death by suffocation can follow in a matter of minutes. The maximum reasonably safe uninterrupted currents are 9 milliamps ac for men of average weight and condition and 6 milliamps for women. At higher values of current, the heart ceases its normal rhythm and results in a condition referred to as ventricular fibrillation. This condition is an uncoordinated asynchronous contraction of the ventricular muscle fibers of the heart, in contrast to the normal coordinated and rhythmic contraction. In this condition, a person may lose consciousness in less than 10 seconds and can have irreversible brain damage within 4–60 minutes. The only way to correct this condition and return the heart to its normal cycle is to apply defibrillation to the chest.

A defibrillator is used to apply an electrical pulse of a safe energy level to the chest to terminate defibrillation. Based on experiments with animals and the results extrapolated to humans, the nominal value of current that may cause ventricular fibrillation is approximated by

$$I = \frac{K}{\sqrt{T}}$$

I is the current in milliamps, K is a constant depending on body weight (K is 165 for a 154-pound man), and T is the time is seconds.

At current values about 5 amperes, the heart and respiratory system may stop functioning. At higher values, severe burns and damage to muscle, tissue, or vital organs can occur. The effect of contact with energized electrical circuits depends on the contact points, the path through the body and the affected organs, the resistance of the path, the magnitude and frequency of the voltage at the contact points, and the duration of the current through the body. In terms of electrical circuit characteristics, the human body is analogous to a circuit having a resistor and capacitor in parallel. The capacitor is very small. Unfortunately, the body is most sensitive to normal 60 hertz current and is less sensitive to dc and higher frequency currents. The dc current level to produce fibrillation is three times the 60 hertz value. The body tissue has a normally high-resistance shell if the skin is not wet, broken, or scratched and a low-resistance core consisting of the blood stream, bone, and body tissue.

Skin resistance has been measured at 30,000–40,000 ohms per square centimeter under thoroughly dry conditions; however, wet skin or scratched skin drops to a fraction of this value. The resistance of dry skin at the point of contact is decreased with contact surface. Above 50 volts, the outer skin high resistance breaks down. Measured minimum resistance values include the following:

Hand-to-hand	Dry	18,000
Hand-to-feet	Dry	13,500
Hand-to-hand	Wet	930
Hand-to-feet	Wet	610

During criminal electrocutions, 8–10 amperes were measured at 2000 volts ac indicating minimum body resistance of 200 ohms. Five hundred ohms are commonly used as minimum resistance. Using 500 ohms, we can see that even 12 volts could produce a dangerous current. Under most circumstances, 50 volts would be considered safe, and the *NEC* recognizes this level in many requirements. Some sources consider 30 volts or less safe. In very wet locations, however, even lower voltages could be dangerous. There have been many electrocutions at "low" voltage; "low" voltage does not imply immunity from the dangers of electrocution.

LIGHTNING PROTECTION FOR CHEMICAL-PROCESS FACILITIES

Lightning strikes have been the cause of numerous safety incidents in chemical-process facilities. According to various accident reports, explosions, power outages, process upsets, and so on, have resulted from lightning storms. These are some of the incidents:

- Lightning struck a storage tank containing 500,000 gallons of methanol. An explosion in the vapor space inside the tank rocketed the 50-foot-diameter tank 75 feet in the air.

- Lightning struck twice at a petrochemical facility and caused extensive damage to process vessels, walls, and piping.
- An underground acrylonitrile storage tank exploded during a lightning storm.
- A dust collector on the top of a structure processing a combustible dust exploded during a lightning storm.
- A unit-operations controller and PLC locked up during a batch sequence step as a result of a lightning strike.

Safety incidents resulting from lightning are of the following types.

1. Explosions of storage tanks or vessels that contain flammable vapors, gases, or dusts. Many of these lightning incidents have occurred worldwide. It is the most serious type of lightning incident because of volume of process material. Large tank farms are deceptively dangerous in this respect.
2. Damage to structures that do not contain flammables, especially brick and nonsteel frame structures like research buildings or cooling towers. Safety reports do not reveal major incidents, but the possibility exists.
3. Electrical power system disturbances and outages with possibly fires and burnouts of substation transformers, switchgear, and polelines. Some processes are extremely sensitive to power system upsets.
4. Disruption and damage to process instrumentation, PLCs, process computers, and other electronic instrumentation and controls. The microelectronic digital circuitry in current lines of instrumentation is far more sensitive than that of previous systems. The increased volume of digital sensors in chemical-process facilities is at risk to lightning and other noise and electrical transients. These transients occur because of lightning even some distance away.

Lightning Storm Physics

The generation of electrical charges in cloud formations during a thunderstorm is a complex phenomenon that results from the interaction of rapid updrafts of air, condensation of moisture particles, and temperature differences within the atmosphere.

The separation of electrical charges results in vast concentrations of usually negative charges in the lower portions of the cloud and positive charges in the upper part of the cloud and on earth. The vast area of charged clouds can produce potentially tens or hundreds of millions of volts between charged areas and earth and oppositely charged areas in clouds. The charge center, fueled by air turbulence, can continue to accumulate charges, increasing the voltage at the cloud until the voltage stress between cloud and earth exceeds the dielectrical strength of air. This dielectrical breakdown can occur between cloud and earth or cloud to cloud. Studies of this phenomenon using specialized high-speed photography reveal the following process.

The initial spark, a pilot streamer, starts its irregular, stepped path to earth by following a line of the concentration of voltage gradient. Pilot streamers proceed downward to earth in successive hesitating steps as long as the charge movement from the cloud to the tip of the streamer sustains the breakdown voltage. If not, the movement ceases and the leader terminates. As the pilot leader approaches the earth, equal and opposite charges on the earth move rapidly in the general area under the leader and eventually produce an upward streamer from the earth. When the streamer contacts the downward pilot, the conductive path is established between the cloud and the earth, and the fully visible lightning strike occurs. Note that most lightning strikes involve multiple strokes, each one following the ionized path of its predecessors.

The pilot streamer does not commit itself to an exact target on earth until it is within the final jump to earth. The length of this final jump is called the striking distance, a term that will be used in defining the protected zone above a tall structure.

The current in a lightning discharge is the source of its destructive power. Current values vary widely. Measured currents on lightning strikes to the Empire State Building indicate that more than 40% of the strikes measured peak values of at least 10,000 amperes and 25% were at least 65,000 amperes. Field data support an average value of 15,000 amperes; the largest recorded value is 345,000 amperes. The magnitude of the current depends on the stored charge in the cloud and the voltage difference between the cloud and the earth.

Most lightning strokes last only microseconds. Typically, the current peaks in 2–10 microseconds, decreases to one-half its peak value in 20–50 microseconds, and is zero within 100–200 microseconds. There are, however, a percentage of strokes that are long duration (in milliseconds), low current (200–1000 amperes), "hot" lightning strokes that can produce severe burning.

The route of a lightning discharge is such that it always seeks the path of least resistance to the opposite electrical charges on earth. It is a fundamental principle of electrostatics that opposite charges exert an attractive force on one another. As a pilot streamer approaches the earth, it has a natural tendency to strike taller objects (towers, poles, etc.). Tall structures have the same electrical charge as earth and project into space. They are the preferred target because of the shorter breakdown distance.

Lightning can strike the same location repeatedly. The top of the Empire State Building has been hit as many as 48 times in 1 year; during one storm, it was struck eight times within 24 minutes.

The instant lightning strikes the top of a building, it raises the point of contact by thousands of volts relative to earth and other parts within that structure. Even metallic parts like storage tanks, piping, and control panels, which are connected to earth via a different path in the structure, may be at thousands of volts relative to the strike area if they are not connected (bonded) to that location.

If the lightning discharge can find a conductive low-impedance path to earth that can tolerate the very high current pulse without blowing apart, overheating, sparking at joints, or producing excessive voltage, it will do so with minimal, if any, damage. Any continuous metal path, like structural steel or a conductor that is adequately in contact with earth, can handle a lightning strike. If lightning strikes a tree, cooling tower, or brick structure, it does not have a direct path to earth and will develop its own path of least impedance, causing damage and burning along the way. Its avowed purpose is to reach earth and neutralize the opposite charge on earth. Lightning protection of buildings and structures, including process structures, tall stacks, and tanks that contain flammable vapors or gases, is described in NFPA Standard 78. The concepts and practices in this standard have been developed and used successfully for decades and are based on the lightning rod concept developed by Benjamin Franklin.

The ultimate goal of these protective systems is to attract and channel lightning strikes to earth without causing damage to people or property. The lightning rod, or air terminal system, does not prevent lightning. A small point discharge of charges flows from air terminal or other tall metallic objects to clouds, but it is usually insufficient to prevent a charge accumulation in a storm cloud. In fact, the lightning protection systems described in this standard attract lightning and thereby prevent objects below them from being struck if they are within the protection zone of the protected structure.

Tanks and other vessels and structures made of at least 3/16-inch thick steel are considered of adequate thickness so they will not be damaged by a lightning strike nor will the metal reach the autoignition of the particular flammable. For materials other than steel and for flammables that have low autoignition temperatures, it would be prudent to calculate temperature rise based on a 15,000 ampere average strike.

Structures that do not provide a continuous metal path to earth can be protected either by providing a system of air terminals connected to earth or by shielding them within the protected zone of an adjacent protected structure. Examples of these types of structures include brick or wood structures, process or control buildings, and cooling towers.

A system of air terminals of the proper length and spacing, connected to down conductors and finally connected to earth at a ground rod, plate, cable, or building rebar, is the protective system described in NFPA 78. The physical arrangement of these elements is very important. It is essential to provide multiple paths from air terminal to earth and to provide as many paths as possible. Multiple paths divide the lightning current to lower values per conductor and provide a close approximation to the Faraday cage effect.

An open steel structure, typical of many chemical-process facilities, can be considered inherently protected by the building steel. Process equipment and piping on the roof of process structures should, however, be at least 3/16-inch thick if it is steel and possibly thicker if it is aluminum or another metal; it should also be bonded to building steel. If metal tanks are on isolated pads or are gasketed, special bonding to building steel may be required to avoid isolated metal sections. Plastic vessels or piping may require a system of air terminals.

Equipment within a structure is subject to a phenomenon called sideflashing. Sideflashing is sparking resulting from differences in voltage between tanks, vessels, piping, control panels, and other metal objects and the down conductors that can occur as a result of lightning currents if the equipment is not bonded to the down conductors. The difference results from inductive coupling and connection to another ground point. A useful rule of thumb is to bond metal objects within 6 feet of an outside column or down conductor to the column or down conductor.

For structures that are not metal and require an air terminal system, several criteria are given in NFPA 78:

1. Air terminals should extend a minimum of 10 inches above the object to be protected. The air terminal diameter should be a minimum of 3/8 inch if the structure is less than 75 feet and 1/2 inch if it is higher.
2. The minimum size down conductor is smaller than a No. 8 copper conductor; however, No. 6 or No. 8 should be used for reliability and mechanical durability in chemical plant environments.
3. At least two paths should exist from every air terminal to ground.
4. Air terminal arrangements and spacing for various roof designs are described in detail.
5. If a down conductor is run through iron conduit tubing or pipe, it must be bonded at both ends.

Iron conduit or tubing develops high impedance in the down conductor and could cause sideflashing. Adequate grounding is essential for proper lightning protection for buildings. Grounding provides metallic contact with earth. Low resistance is highly desirable. However, NFPA 78 does not specify a minimum value; some companies, however, require 5 ohms minimum.

The distribution of ground terminals is very important. Lightning should be able to find a direct and easy path to earth no matter where it might strike the top of a structure. This is why there are always at least two down conductors and the down conductor path to earth must be as direct as possible, avoiding bends and not running through a conduit (unless it is bonded to it). Each down conductor shall terminate in a ground terminal, and the size, length, and depth must be commensurate with soil conditions. In rocky conditions that offer high resistivity, a loop conductor encircling the structure may be required to interconnect ground terminals. A ground terminal may be a ground rod, ground plate, or bare copper conductor. Ground terminals shall not be more than 60 feet apart around the perimeter of the building. Tall stacks require at least two terminals.

A concept that has been used in the chemical industry for building lightning and computer grounding is the Ufer ground. This system uses the foundation rebar in buildings as a ground mat. Ufer, an engineer, found that reinforcing rods in concrete are as effective as a ground rod in earth; numerous tests have been performed that support his conclusion.

The advantage of the Ufer ground is the size and extent of the rebar grid. The building's structural steel columns are bonded to the foundation rebar using the anchor bolts on the column (Figure 6-1), and the rebar is tied together. The Ufer ground is only possible on new facilities where the connections to the rebar can be inspected. Underground corrosion, especially cathodic protection systems, need to be coordinated with electrical earth grounding to ensure compatibility and code compliance. Tall structures, like towers, poles, and stacks, can protect lower structures if they are within the taller structures' zone of protection. The zone of protection is the area in which the probability of a strike occurring is very low because the preferred target for a lightning strike is the taller protected structure or earth next to the protected zone.

Until recently, the zone of protection was considered to be a cone with its apex at the top of the protected structure and its sides at a 30-degree angle to the vertical. This concept did not explain why tall structures received strikes on the side. A study of strikes on the sides of tall power transmission towers resulted in the development of the "rolling sphere" concept of the protected zone (Figure 6-2). This concept explains why structures could be hit above 100 feet and why structures above 100 feet provide no additional protection. The protected zone is the space under the surface

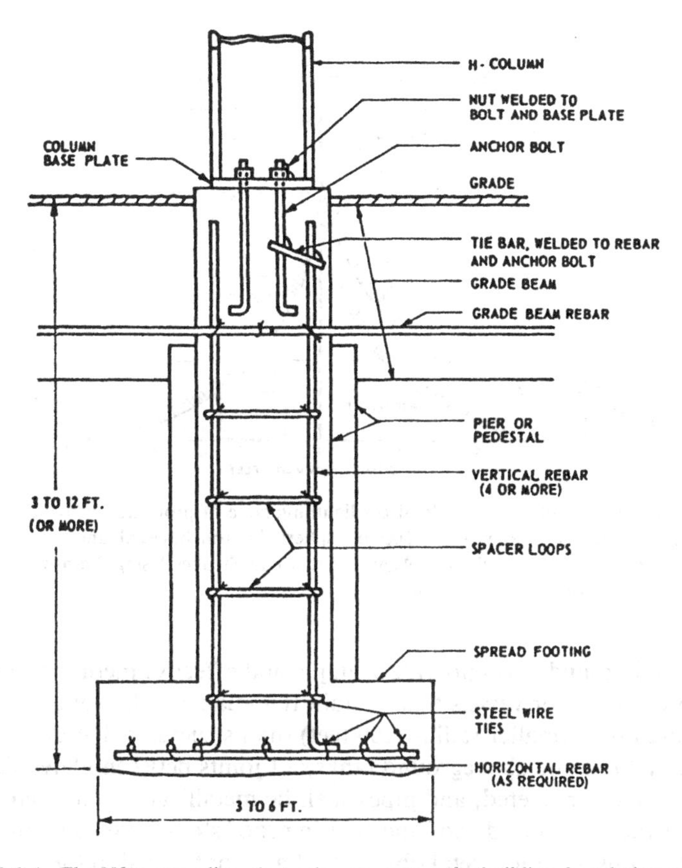

FIGURE 6–1. The Ufer grounding electrode system uses the building foundation rebar as a grounding electrode for lightning protection and computer grounding. The system is particularly effective in poor soil conditions as discussed in the Fagan and Lee (1970). (Fagan, J., and R. H. Lee, The Use of Concrete-Enclosed Reinforcing Rods as Grounding Electrodes. *IEEE Transactions on Industry and General Applications* IGA-6, No. 4, July/Aug., 1970, p. 338, Fig. 1.) © 1970 IEEE.

of a sphere, with a radius based on the striking distance tangent to the earth and resting on the taller protected object. The striking distance is the final step in the stroke where the point to be hit is determined. It is the radius of the sphere that determines the protected zone.

A simplification of the rolling sphere concept is used for structures less than 50 feet high. Structures that contain flammable vapors and gases must be tightly sealed to prevent the escape of gases or vapors. Openings must be protected against the entrance of flame and isolation.

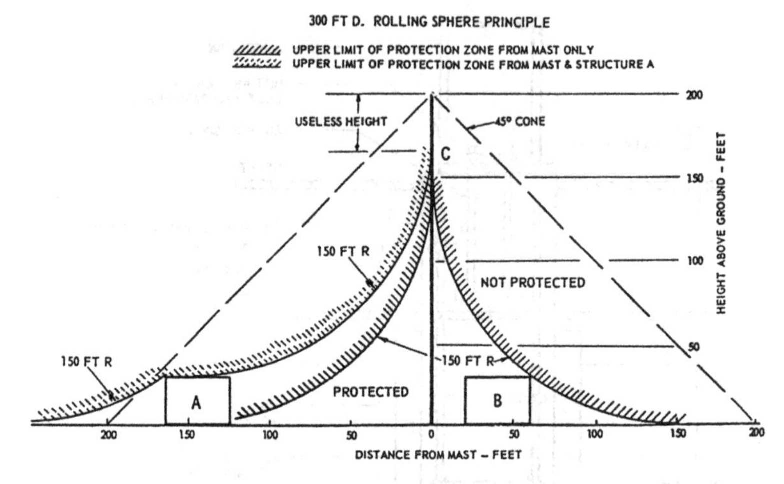

FIGURE 6-2. Rolling sphere principle of the lightning zone of protection provided by a tall mast. The protected area is external to a 300 foot sphere that touches earth and the mast. (Lee, R. H., Protection Zone for Buildings Against Lightning Strokes Using Transmission Line Protection Practice. IAS 77 Annual, Fig. 4.)

Overhead ground wires provide a simple and effective method to protect large storage tanks or vessels that contain flammables. The zone of protection is based on a smaller radius (100 feet) than structures that do not contain flammables. For aboveground tanks, all joints between plates shall be welded, bolted, or riveted, and pipes shall be metallically connected where they enter the tank. Fixed roof and floating roof storage tanks require special precautions to avoid metal objects that are not bonded to the tank shell and to minimize the accumulation of vapors at seals.

Some explosions in flammable storage tanks have been attributed to sparking at a cable during an electrical storm. It was suspected that the cable received an induced charge from a lightning strike and flashed over inside the tank. NFPA 78 provides the necessary safety criteria for most situations, but additional protection may be required where the frequency and severity of storms are high or where large volumes of flammable or toxic materials are stored. For instance, an additional overhead ground wire or added air terminals at an open vent may be inexpensive insurance.

Air terminals do not prevent lightning strikes; the small leakage current from one air terminal or even multiple air terminals to a cloud is insufficient to prevent the accumulation of static charge in a cloud. There is, however, a specialized air terminal system, called a Dissipation Array, that is designed to "bleed off" charges in clouds and prevent lightning strikes. The system consists of an ionizer, ground current collector, and service connections. The ionizer is a space charge generator based on multipoint discharge. Point conductors subject to a high voltage will emit a small ionization current.

Early studies of lightning rods indicated the necessity of eliminating all charge accumulation to prevent a lightning strike, but proponents of the Dissipation Array indicate this is not true. The ionizer is located, oriented, designed, and elevated in such a way as to take advantage of the electromagnetic field produced by charged clouds. The size and location of the ground current collector must be such that the charge is collected and provided with a preferred path. The current in these paths is usually less than one ampere. System connections are provided from the ionizer to points of collection. The ionizer can be a roof array, umbrella, or conic section (Figure 6–3).

The Dissipation Array concept is somewhat controversial. It is not a part of NFPA 78, and therefore each user must evaluate its applicability to individual situations. It has been applied to TV and radio towers and other facilities.

Lightning Protection of Electrical Power Systems in Chemical-Process Facilities

Lightning strikes can produce severe overvoltages on electrical power lines and equipment, either by direct strokes or by induced power surges on power lines. The results can be a power blip, a minor outage, or, if distribution equipment is damaged, a major outage. Lightning overvoltages can cause fire and explosions in transformers and switch gear. Power system disturbances can cause safety problems for most processes. The type of process and the electrical system design determine the response of the unit to this upset.

The protection of electrical power systems against lightning overvoltages is accomplished by equipment and wiring design and by protective devices like surge arresters and capacitors. The Institute of Electrical and Electronic Engineers (IEEE) and other ANSI standards cover the design and testing of substation transformers, switch gear, arresters, grounding, and power cable ratings. These standards involve the process of correlating the electrical insulation strength of electrical equipment with anticipated overvoltages and the characteristics of surge protective devices. A surge arrester is a protective device for limiting the magnitude of the surge voltages at transformers, motors, and switch gear. Choices as to the rating and application of these devices involve higher initial cost versus long-term reliability. In chemical plant operations, reliability and safety should not be compromised.

FIGURE 6–3. Lightning Strike Prevention (Dissipation Array System) in a process facility. The umbrella-like devices on the top of the off gas stacks are called ionizers and are intended to dissipate electrostatic charges from a protected facility created by a storm charged cloud. (Lightning Eliminators and Consultants, Inc., Boulder, Colo.)

Most in-plant electrical systems are reasonably well protected against lightning overvoltages. The power lines are short relative to utility lines; therefore, the exposure is lower. Tall structures, like towers and stacks, provide a zone of protection for some parts of the electrical system. Cabling for inplant facilities is in the cable tray or conduit or in rings and saddles on messinger cables between poles. These facilities shield the power wiring from the effects of lightning discharges.

Plant electrical utility systems have a greater exposure to lightning. The power lines are longer and higher in the air. Most power lines are protected by an overhead "static" wire intended to shield the power lines; however, the protection is not perfect.

Chemical-process facilities, therefore, are vulnerable to the effects of lightning storms in the power utility system. For this and other reasons, backup power systems are an alternative to power interruptions for sensitive process units.

Lightning discharges can also cause serious safety problems with electronic instrumentation, PLC, and process computer systems. A lightning discharge can cause differences in voltages in structures, which can introduce noise signals and damage to microcircuits. The new generation of microcircuitry is vulnerable to these types of low-level electrical disturbances. The microcircuit boards are smaller and have less spacing than previous systems; therefore, they are very susceptible to overvoltage failures. Overvoltage failure is the major cause of electronic component failure. The failure can be immediate, or the weakened parts (dielectrically weakened) may not fail until later, perhaps at a critical time. Lightning also radiates high-frequency energy. Observe the static on a radio during an electrical storm. A lightning stroke acts as an antenna, propagating electromagnetic fields in all directions for many miles. The wiring to instrumentation can act as an antenna to pickup for these fields. A transmitter at the top of a tall structure is especially vulnerable. Even underground cable can be damaged by lightning discharge earth currents.

The lightning protection measures used to protect buildings and electrical systems are essential, but additional measures are required to protect electronic instrumentation. They include

Grounding for instrumentation
Shielding of wiring and enclosures
Twisted pair wiring
Electronic overvoltage protection

STATIC ELECTRICITY AS AN IGNITION SOURCE

Static electricity is generated by the physical contact and separation of materials and objects and is a fundamental part of chemical processes wherever gases, liquids, dusts, or solids flow or move in pipes, ductwork, tank trucks, loading drums, or vessels. Under a limited range of conditions, static charges can accumulate and build up voltage to the point where an electrical spark results; such as spark can ignite a flammable gas, vapor, or dust.

Static electricity can have other, less catastrophic but still undesirable, effects. For example, static sparks generated by operators can result in discomfort and possible injury. Therefore, operator complaints of static should prompt static safety audits if flammable materials are involved. In some process situations, static electricity can cause, contamination of process materials. Static electricity generated by personnel can cause damage and ultimately failure in critical process control microcircuits.

Static electricity safety has been studied for a number of years in various industries, especially in the petrochemical industry, where the consequences of static explosions in large storage tanks of highly flammable gases and vapors have been serious. "Unexplained" explosions in refineries prompted the well-known study by the Royal Dutch Shell Laboratory (Klinkenberg and Van Der Minne). The API has developed a recommended practice, API RP 2003, *Recommended Practice for the Protection Against Ignitions Arising Out of Static, Lightning, and Stray Currents* (1982). This standard was developed to provide guidance for static safety in refineries, but it also applies to chemical processes. It was revised in 1982 to provide additional guidance on acceptable maximum flow rates of liquids in pipes to minimize static electricity generation. Static electricity safety considerations for other processes, including those involving flammable dusts, fibers, gases, and certain industrial and commercial processes, are in NFPA 77 (1983), *Recommended Practice on Static Electricity.*

Dusts, particularly chemical and plastic dusts, plastic pellets, and plastic sheet-line processes have a strong capacity to generate and retain large static charges, which can ignite any flammable dusts or vapors that may be present. Explosion investigation teams sometimes name static electricity as the ignition source when no other possible source could be found. Using the principles we shall discuss and, in some situations, doing tests that it may be possible to gain a higher confidence level that static electricity could have been the ignition source. It is equally important to know when it could not have been the ignition source so the search for the real source can continue.

Figure 6–4 illustrates a process situation that can generate static electricity. In this and similar process situations, process flow rates, conductivity of the process fluid, metal versus plastic piping and containers, and agitation rates determine static charge accumulation. Static safety distinguishes itself from hazardous area safety because the spark can occur inside process equipment.

Fundamentals of Electrostatics

The normal state of atoms in matter is electrically neutral. The negative charge of the outer mobile electrons is balanced by the positive charge of the nucleus. When electrons move from their normal state to another location, negative charges are created in the region to which they move, leaving a region of positive charges.

Wherever there are areas of electrical charge, electrostatic forces exist. When the charges are unlike, that is, negative and positive, the forces attract the charges to each other. When the charges are like, the forces repel the charges. Charges generated by static electricity always occur in pairs.

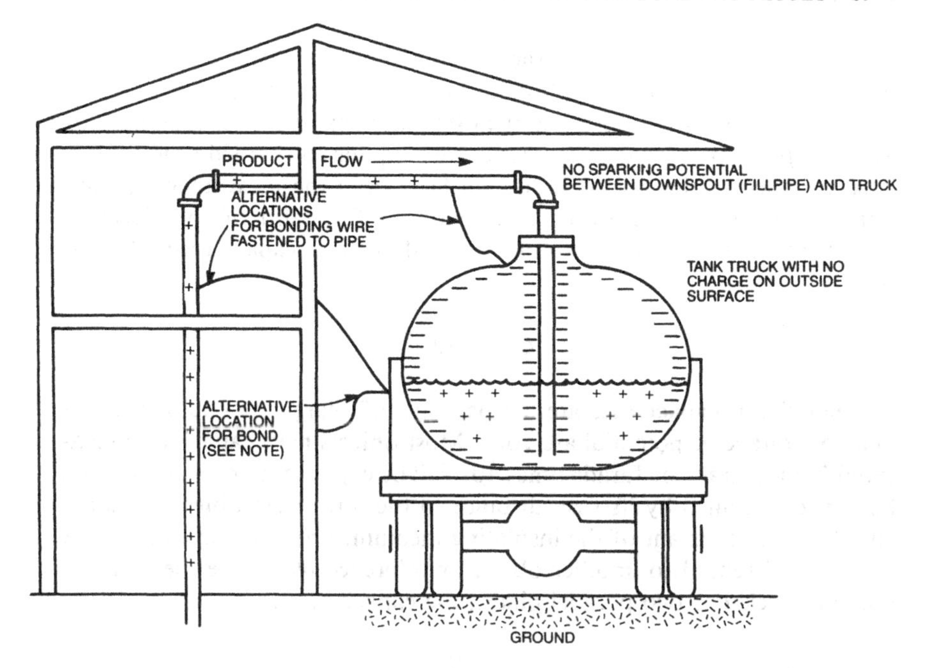

NOTE: The alternative bond location to the steel rack is permissible if it is inherently connected with the loading piping.

FIGURE 6-4. A process situation, loading of a tank truck, will generate static electricity due to the flow of liquid and the electrical bonding to eliminate voltage differences. (American Petroleum Institute, API RP2003, 5, Fig. 2.)

There is also a potential or voltage difference between charges, which is a measure of the electrostatic forces present. In electrostatics, the voltage levels are in thousands of volts, or kilovolts. For instance, a person walking across a carpeted floor can generate approximately 5 kilovolts.

An electrical force field exists between charged areas, which is represented by lines of force. The density of these lines of force is called the potential gradient or electric field intensity. The force field and potential gradient vary with the radius of the curve of the electrodes and the distance between them. Sharp points can produce high stress, causing voltage breakdown to occur more easily. For instance, needle points 1/2 inch apart will spark over at approximately 14 kilovolts and spheres 1/2 inch apart would require 20 kilovolts. This phenomenon explains why sharp points inside storage tanks can be spark promoters to a charged liquid.

The above discussion pertains to air as an insulator, but ionization can also occur along surfaces between charged objects. Surface conditions, in particular, humidity can cause tracking and flashover along a surface. The surface material can be susceptible to high humidity conditions.

Isolated conductors in an electrical field are charged by induction; that is, they "pick up" a charge via capacitive coupling to charged objects. Isolated conductors charged by induction have been the most likely ignition source in a number of static investigations. The ratio of the electrical charge, Q, on a body to the voltage, V, is defined as capacitance, C. That is,

$$C = VQ.$$

The units of capacitance are farads, units of charge are coulombs, and units of voltage or potential are volts. Most objects have capacitance measured in picofarads or farads times 10^{-12}. The capacitance of a conducting body is determined by its size, distance to the referenced object or plane, and dielectric constant of the insulating medium. Capacitance and voltage must be referenced to another object or reference plane like the earth or building steel. For two parallel plates, the capacitance is

$$C = \frac{KAE}{t}$$

where K is a constant, A is the area of the plates, t is the distance between the plates, and E is the relative dielectric constant of the insulation between the plates. E is 1 for a vacuum and has the following approximate values for some chemicals:

	E
Benezine	2.3
Toluene	2.4
Xylene	2.4
Methanol	33.7
Acetone	21

The capacitance of various objects used in electrostatics are in picofarads; some approximate values are as follows:

	Capacitance (picofarads)
Human being	200
Automobile	500
Tank trunk	1000

Capacitance is the source of stored electrical energy in an electric field that can provide an igniting spark. The stored energy in the field is

$$W = 0.5 \times C \times V^2$$

where W is the energy in microjoules, C is the capacitance in picofarads, and V is the potential in kilovolts. W is the maximum energy available for a spark. The actual energy in the spark is lower because of the resistive losses in the circuit and at the contact points. Sparks produced by a conductor, such as an isolated section of metal pipe, can contribute the total stored charge because of the inherent high conductivity. Insulators such as a plastic sheeting, however, restrict stored charge flow from their surfaces by their low conductivity. Therefore, isolated conductors present greater potential for spark ignition.

Generation and Accumulation of Static Charges

Static electricity is generated at the surfaces of different materials as they contact and separate. This surface phenomenon results from the migration and separation of electrons. The rate of generation increases with the flow rate of the two surfaces and the contact area.

For liquids flowing in a pipe, a fixed layer of charges remains on the pipe wall while a mobile layer of charges remains in the liquid close to the wall. These two charged areas are called the electrical double layer and can be viewed as the plates of a charged capacitor. The liquid in motion carries the charged particles through the piping system into a storage tank or vessel. The oppositely charged particles on the piping wall flow to earth. The flowing liquid carrying the charge is referred to as the streaming current.

Generated charges have a natural tendency to reunite if a conductive path is present. If the charged objects are isolated, they will remain charged and will continue to accumulate charges as long as there is flow. The voltage difference between the charged objects, however, forces a leakage current to flow between them. Equilibrium occurs when the leakage rate equals the charging rate.

In the case of liquids flowing in a metallic piping or vessel system, the only resistance to charges recombining is the fluid itself. Flows that have low resistivity or high conductivity allow charges to equalize quickly, and charge accumulation is minimized. Therefore, liquid conductivity is a key issue in static safety. The conductivity of a liquid is a measure of its ability to accumulate static charge.

Conductivity is the inverse of resistivity. Resistivity is measured in units of ohm-centimeters or ohm-meters; conductivity is measured in units of mhos per centimeters or meters. A unit of conductivity is a siemen, or one mho per centimeter. API RP 2003 (1982) indicates that liquids having conductivities greater than 50 picosiemens per meter do not usually have significant charge accumulation.

The Royal Dutch Shell Research and Development report on electrostatics provides the following insight concerning refinery operations:

- Static electrification is a function of the type and concentration of some trace compounds present in oil products.
- Certain metal salts, when used in combination with other metal salts, can greatly increase conductivity.
- Combinations of salts are significantly more effective than individual salts.
- The presence of free water in a petroleum product significantly increases electrification.

When a charged liquid enters a storage tank, truck, or vessel, time is needed for the charges to migrate to the side of the tank. This is referred to as an equalization process, and time for this to take place is called the relaxation time. The relaxation half-time, or the time for an electrical charge in the liquid to decay to one-half its initial value, is calculated by

$$T = \frac{E \times 6.15}{C},$$

where

T is the time in seconds for the magnitude of the charge to decay to one-half its initial value
E is the relative dielectrical constant
C is the conductivity of the liquid in the picomhos per meter

The half-time does not depend on the size or dimensions of the vessel but is solely dependent on the electrical characteristics of the liquid. The longer the relaxation time, the greater the exposure to static charging and the greater the potential for static ignition. This is why it is necessary to wait a period of time before inserting level gages or other devices into a tank after transfer is complete. One rule of thumb suggests that the risk is minimal if the half-time is less than 0.01 seconds. The relaxation time for toluene is 15 seconds; for methanol it is 9 microseconds. Xylene has a dielectric constant of 2.38 and a conductivity of 0.1 picosiemens per meter. Therefore, the half-time for xylene is 146 seconds.

Charge neutralization assumes that the piping and vessels are metallic and are connected together (bonded) and connected to earth (grounded). In situations in which plastic piping and vessels or metallic sections are isolated by piping, static charge accumulation continues unabated, and the risks of static ignition are higher.

Static Ignition

For static ignition to occur, the following conditions must be present:

- Electrostatic charge generation must occur.
- There must be the means for static charges to accumulate.
- The electrical field resulting from the static charge must be strong enough to produce a spark of sufficient energy to cause ignition.
- An ignitable mixture of vapor–air or dust–air must be present.

The ignition energy of flammable gases, vapors, and dusts has been determined for a number of materials based on standard testing. Values for some vapors and gases are as follows:

	Minimum Ignition Energy (millijoules)
Acetone	1.15
Acrolein	0.13
Toluene	0.24
Xylene	0.2
Ethyle acetate	1.42

These values depend on test methodology. The tested minimum ignition energy depends on the concentration of the flammable. These reported values represent low numbers relative to actual conditions, but they could be used to determine if ignition is possible.

Preventing Electrostatic Ignition

The prevention of electrostatic ignition consists of controlling or eliminating one or more of the factors or conditions that can lead to static ignition. That is,

- Reduce static charge generation.
- Limit static charge accumulation.
- Eliminate potential static spark promoters.

- Prevent flammable gas, vapor, or combustible dust from occurring in a volume within the flammable limits.

Reduce Static Charge Generation

For liquid systems, static charge generation can be reduced by the following:

- Reduce flow rates.
- Minimize agitation and turbulence.
- Minimize droplets of water and other particulates from settling through liquid.
- Avoid pumping in air.
- Avoid splash filling.
- Use bottom loading whenever possible.

The objective of all these actions is the reduction of surface area contact.

Maintenance of the flow rate below acceptable limits has been recognized in the petroleum industry as necessary to limit electrostatic charge generation. The 1967 edition of API RP 2003 indicated that a maximum flow velocity of 15–20 feet per second would keep static charge generation within acceptable limits. Therefore, during filling and unloading of tank trucks, cars, or other similar operations, flow rates were to be limited. The 1982 edition of API RP 2003 indicates, however, that based on current data, bulk loading rate should be used to determine the maximum acceptable flow rate. It recommends that the maximum linear flow velocity be determined by this formula:

$$V \times d \leq 0.5,$$

where

V = flow velocity in meters per second
d = pipe diameter in meters

In addition, V should never exceed 7.

API RP2003 indicates flow rates for selected pipe sizes. This standard contains a number of recommendations and comments, including the following:

- Impurities, like water, metal oxide, and chemicals, can increase static charge generation.
- Filters are significant charge generators, therefore pipe length down-

stream of a filter should be installed to provide least 30 seconds of relaxation time before the charged liquid enters a tank truck, storage tank, or other vessels where flammables vapors of gases may be present in the vapor–air space.

- Splash filling and loading will contribute to electrical charge generation. Fill pipes should be as near as possible to the bottom of tanks. Bottom loading is preferred.
- The settling of minute quantities of fine water droplets, iron scale, and other particles can contribute to static charge generation.

Limit Static Charge Accumulation

Static charge accumulation can be limited by promoting charge equalization. If the charged areas are connected via metal piping and vessels, charge equalization occurs through the liquid. The methods of improving equalization are as follows:

- Improve conductivity of fluid.
- Provide metallic piping and vessel systems that are bonded together and connected to the earth or building steel.
- Humidify the area.

The electrical charge generated in a liquid will move to an opposite charge on the tank or piping wall at a rate determined by the conductivity and dielectrical constant of the liquid. The conductivity of certain liquids can be significantly increased by the addition of antistatic additives if the additives are compatible with the process. The Royal Dutch Shell Research and Development report indicates the profound effect of trace compounds and metal salts on the conductivity of petroleum fluids. Fluids with long relaxation times due to low conductivity require special attention.

Grounding and bonding are essential to minimize static charge accumulation. Bonding is the joining together of metallic bodies to eliminate any voltage differences that might occur as the result of static charges on one of the bodies. Nonconductors like plastics cannot be bonded together or grounded to achieve voltage equalization because of the immobility of charges in nonconductors. Grounding is the connection of metallic parts to earth or some other potential reference plane like building steel to achieve voltage equalization.

Grounding and bonding for static charge equalization differ from grounding for lightning and electrical equipment because for the former the voltages are high (kilovolts) and the currents are low (microamperes). Therefore, low resistance and current carrying capability are not requirements for static grounding and bonding. A resistance as high as one megohm will equalize voltages. Any reliable conducting connection is acceptable, but it must be able to endure the chemical, corrosive, and mechanical service conditions.

Static bonding and grounding cables are sized for the environmental service conditions. The electrical grounding systems used in buildings for lightning protection and electrical system grounding provide an effective static grounding system. Additional bonding and grounding may be required to process piping, tank trucks, drums, and other metallic objects that may be isolated and not connected to the static generating piping.

All metallic parts of the piping and vessel system flow must be reliably connected together. Figure 6-4 shows typical bonding connections for top loading a tank truck. It is especially important to bond the top dome to the fill stem. Bonding connections should be made before the dome cover is open and should be in place until the transfer is complete and the dome is closed.

Bonding connections should be inspected for damage and corrosion and tested for continuity. Instrumentation for checking ground continuity is available. The output signal, indicating the absence of ground continuity, can alarm or shut down transfer pumps.

Bond wires are not usually required around flexible, swivel, or sliding joints or at metallic pipe flanges. Normal metallic contact is usually sufficient to minimize static. If a visual inspection indicates corrosive or other conditions that may prevent continuity, however, a simple continuity check may be necessary and bonding jumpers may need to be added. A conductive path from the drum or can to the fill line must exist to minimize any static potential at the fill point. The containers are inherently bonded to the fill system if both are on a metal platform.

In situations in which the piping, ductwork, tanks, or vessels are nonconductive, electrical charges cannot be neutralized by bonding or grounding. Static charge accumulation will continue to a level dependent on the charge leakage rate. Note that an electrical field exists outside plastic piping. Usually, however, nonconductive surfaces do not produce ignition capable sparks. An isolated ungrounded metal object, for instance a metal hose or pipe section isolated by nonconductive piping or fittings, can, however, accumulate a significant static charge by induction and spark over to any nearby grounded object.

Humidification can limit static charge accumulation by providing a conductive film on any insulating surface. This film provides a leakage path for static charge accumulation. When the relative humidity is high, the potential for static sparking is less.

Eliminate Potential Static Spark Promoters

Any conductor in a storage tank or other vessel in the flammable vapor space that is not bonded to the tank and fill piping can be charged by induction and possibly spark to the grounded tank. Level sensors are possible candidates as spark promoters.

Prevent Occurrence of Flammable Gas or Vapor

In order for a flammable gas or vapor to propagate a flame and explode, the vapor or gas mixture must be at a concentration within the flammable limits. For liquids that produce flammable vapors must be at a temperature above their flash point. to produce a flammable air–vapor mixture, low vapor pressure, high flash-point (above 100°F) liquids do not produce flammable vapors unless heated above their flash point. High vapor pressure, low flash-point liquids (below 100°F) usually do not produce vapor spaces above the flammable limit. Intermediate vapor pressure, low flash-point liquids may create flammable concentrations in the vapor space. Purging the vapor space with an inert gas can reduce the concentration below the lower flammable limit, but the addition of inert gas must be designed to be compatible with the vessel's fill and pump out systems.

Clothing and Personnel Electrification

The human body is a conductor. Therefore, if it is sufficiently insulated from ground, it can accumulate and hold an electrical charge. Static charges can accumulate on an outer garment, especially at low relative humidity, if the garment is rubbed against a seat or taken off. If the person is insulated by his or her shoes (leather shoes do not provide insulation; rubber and composition shoes do), the induced charge can produce an incendiary spark for vapors or gases. Precaution includes grounding the person before handling or touching flammables and avoiding isolated conductors, such as hoses or funnels, which could spark to a person.

Dusts

Chemical dusts, particularly plastic dusts, can produce large static changes due to their inherently high resistivity. The minimum ignition energy of dusts is an order of magnitude more than flammable vapors or gases and is dependent on the particular material and the type of test. The minimum ignition energy can, however, be within the capability of the static sparks generated by dusts.

Dusts are produced in chemical processes as by-products or fines from various operations, for example, pneumatic conveying of plastic pellets and blending. During the processing, combustible dust clouds can occur inside or outside process equipment and be ignited by a static spark. As with vapors the isolated conductor is the usual ignition source. Nonconductive piping vessels and ductwork can produce incredible static charges in plastic and chemical dusts because they do not provide a ground path, as metalic grounded systems do.

There also have been a number of explosions resulting from static sparks generated by dusts igniting flammable vapors.

Dust change levels can be measured by the Faraday pail technique and the charge level in millijoules or coulombs per kilogram. In this test the dust is allowed to drop through pipes simulating the actual process piping and then accumulated in a isolated pail. The material is then discharged. Test results can be compared to a dust of known explosion hazard.

Dust static safety is necessary, following the same practices and principles described for vapors and gases.

Gases

Pure gases do not usually generate charges. Gases that contain particulate matter can, however, generate static charges that can accumulate on isolated objects. Carbon dioxide is a particularly static-prone gas and should be avoided if flammables are present.

PROTECTION OF ELECTRICAL SYSTEMS

Environmental Protection

Chemical processes impose severe environmental conditions on electrical equipment and wiring. Rain, dust, humidity, hosedown, submersion, vibration, temperature extremes, mechanical damage, and chemical corrosion have caused wiring and component failures. In fact, environmental factors are a major cause of failures. Water inside electrical enclosures and chemical corrosion are the worst conditions. Water inside enclosures can cause wiring failures or electrical contact or component corrosion. Chemical corrosion can destroy metallic conduits and enclosures.

Environmental conditions should always be considered when selecting enclosures, equipment, and installation designs. The following are examples of actual situations:

An explosionproof limit switch was provided on 200 on-off batch control valves. The enclosure design for the limit switch consisted of a sintered metal cover, which acted as a flame arrester and enabled the switch to acquire its explosionproof listing. The cover, however, easily permitted water to enter the inside of the switch and bridge between the wiring terminals and the metal enclosure. This situation was not discovered until checkout, when the fire sprinkler system in the unit was tested. By then, water had entered a number of switches and cause the entire control system to shut down. The limit switches had to be replaced with explosionproof and watertight switches.

Small traces of a seemingly harmless chemical vapor entered the ventilation system in a process computer room. The vapor coated the printed circuit board connectors with an insulating film, which resulted in intermittent interlock failures. During hosedown of a facility, water accumulated inside at the bottom of a fluorescent lighting fixture and caused arcing to occur across the terminals at the tube pin connector. The arcing eventually caused the plastic light cover to ignite, which resulted in a fire.

The location of electrical equipment and wiring should be such that the exposure to leaks, hosedown, vibration, and so on, is minimized. Out of necessity, however, process instrumentation or other equipment must be located in the process unit. The most effective solution, therefore, is to provide an enclosure that protects against the environment.

Enclosures

Use the terminology in NEMA standard 250, *Enclosures for Electrical Equipment,* to select electrical enclosures that are watertight, dusttight, and hosedown resistant. The terminology defined in this standard (NEMA 3, NEMA 4, etc.) is used on a daily basis in the industry to communicate user needs to the suppliers. For instance, if a pressure switch is to be located in an unclassified, hosedown area, NEMA 250 indicates that an NEMA 4 enclosure should be specified. Table 6-2 lists the NEMA enclosure definitions, the intended use, and the degree of protection.

NEMA 250 does not cover motors or rotating equipment or electrical power equipment operating at more than 1000 volts, nor does it describe the structural features of each type of enclosure. Each manufacturer can meet the functional requirements by its own design.

Each type of enclosure should be capable of passing the following environmental tests:

Rod entry test
Drip test
Rain
Dust
  Outdoor dust—Dust blast; hose method
  Indoor dust—Circulating dust
External Icing
Hosedown
Rust resistance
Corrosion protection

TABLE 6-2. NEMA Enclosure Definitions

NEMA Type	Intended Use	Protection Against
1	Indoor	Contact with enclosed equipment
2	Indoor	Limited amounts of falling dirt and water
3	Outdoor	Wind-blown dust, sleet, rain, ice formation
3R	Outdoor	Rain, sleet, ice
3S	Outdoor	Wind-blown dust, sleet, rain, operation of external mechanisms when ice laden
4	Indoor/outdoor	Wind-blown dust, rain, and hose-directed water
4X	Indoor/outdoor	Same as NEMA 4 and corrosion
5	Indoor	Settling dust, dirt, dripping noncorrosive liquids
6	Indoor/outdoor	Occasional submersion in water
6P	Indoor/outdoor	Prolonged submersion in water
7	Hazardous (classified) locations—indoor	Class I Groups A, B, C, or D
8	Hazardous (classified) locations—indoor/outdoor	Class I Groups A, B, C, or D
9	Hazardous (classified) locations—indoor	Class II Groups E, F, or G
10	Applicable to mines	
11	Indoor	Oil immersion and corrosion
12	Indoor	Dust, dirt, liquids
12K	Indoor	Dust, dirt, liquids, except at knockouts
13	Indoor	Dust, spraying water/oil

Submersion
Air-pressure test
Oil-exclusion test

Each NEMA type enclosure is designed to pass one or more of these tests; however, each enclosure is not actually tested. That is, an NEMA 4 enclosure is designed to pass the external icing, hosedown, and corrosion test; an NEMA 3 is designed to pass the rain, outdoor dust, external icing, and corrosion protection tests. NEMA 3R is similar to NEMA 3, except it is not designed to pass the outdoor dust test. NEMA 3R can be ventilated or nonventilated. This is also true for NEMA 1 and 2 enclosures. All other enclosure types are nonventilated.

Ventilated enclosures are less desirable than nonventilated enclosures because the internal equipment and wiring are exposed to the outside air, which may be humid or have chemical gases or vapors. Ventilation may be required because the internal components, like transformers and power semiconductors, generate heat and require cooling air to maintain their temperature below acceptable values for the equipment. If the ventilated enclosure is exposed to fumes or humidity, it may be desirable to provide a ventilation system with the air pickup point in a "clean" location. If the enclosure houses electronic equipment, air conditioners can be provided to cool the internal equipment in an internal circulation system that does not contact outside air. An external circulation loop provides cooling with the outside air.

NEMA types 7, 8, and 9 enclosures are listed for hazardous locations and comply with UL standards. The appropriate class, group, and T number must be specified for the particular flammable involved.

Explosionproof enclosures are not necessarily watertight or dust tight. The covers on explosionproof equipment must quench the flame developed by an internal ignition and prevent ignition from occurring outside and enclosure. The covers use ground surface or threaded construction at the contact surface, which does not necessarily provide a seal against outside water.

Watertight-or hosedown-protected equipment (NEMA 4 enclosures) have a gasketed cover to prevent water from entering the enclosure, but water leakage has been a problem with explosionproof equipment. Water can enter the equipment and cause the equipment to fail. In recent years, the manufacturers of explosionproof equipment have made great improvements in providing equipment that is both watertight and explosionproof. These designs use an O ring or gasketed seal that does not interfere with the designed flame path.

Purged and pressurized enclosures, complying with ISA Standard S12.4 and NFPA Standard 496, can be used in hazardous locations to provide environmental protection. The internal pressure of the enclosure prevents humidity, dust, rain, and so on, from entering and enclosure.

Enclosures can be constructed of metal or plastic materials. The metallic enclosures include those constructed of cold rolled, galvanized, and stainless steel and of aluminum. Plastic materials include fiberglass, polycarbonate, and polyester. Enclosure manufacturers provide data indicating the acceptability of their enclosure materials with various types of chemicals and corrosives. The NEMA type 4X enclosure is corrosion resistant based on testing in a salt spray environment.

The acceptability of an enclosure for the particular corrosive material in a given application needs to be verified with the enclosure manufacturer. In some cases, samples of the enclosure material can be tested in the particular environment. The type and material of the enclosure door or cover gasketing is very important. If the gasketing fails, the enclosure may no longer be watertight or dust tight. The gasket should be securely attached to the enclosure and constructed of materials that do not erode from corrosive vapors, age, or mechanical damage. Gasket materials include neoprene rubber, urethane elastomer, and silicone rubber.

Enclosures protect people against accidental contact with electrical components as well as against rain, hosedown, dust, and other external conditions, but they cannot protect against condensation or corrosion or conditions that occur because the enclosure breathes. All enclosures breathe to some degree, even with the best cover gaskets, and during ambient temperature variation, condensation can occur inside the enclosure, especially in very humid and Gulf Coast locations.

In these high-humidity areas purged and pressurized enclosures have a distinct advantage, since the pressurized air will keep humid air outside the enclosure. Also, plastic enclosures are less susceptible to condensation and usually are more corrosion resistant than metallic enclosures. They may not, however, be as strong mechanically, nor do they provide protection against electricomagnetic interference (EMI). Electrical grounding in plastic enclosures requires special care, whereas metal enclosures can easily provide grounding.

Enclosures also need to be protected against the entrance of liquids and gases through conduits or uncovered conduit openings in the enclosures. All openings for conduits should be sealed to minimize the possibility of liquids or gases entering an enclosure. Conduits should enter enclosures from the bottom or minimize the possibility of fluids entering.

There are war stories of situations where water or process fluids entered enclosures through conduits. In one situation, a flammable vapor entered a panel when an electrician was measuring voltage using a solenoid-type voltage tester. As the solenoid tester was being removed from an energized terminal, a spark ignited the flammable vapor and caused a flash fire in the panel.

Electrical enclosures must also be protected against mechanical damage. (Note that NEMA 250 does not specify mechanical strength of the types of enclosure.) They should not be in location where they are exposed to damage from forklift trucks or other moving equipment or where piping or equipment must be removed.

The types of electrical enclosures used for electrical motors are defined in NEMA Standard MG-1. Most of the motors used in chemical plants are in totally enclosed fan-cooled (TEFC) enclosures rated for chemical service (Figure 6–5). The stator winding rotor, bearings, and rotor shaft are totally enclosed, with only each end of the shaft and the junction box extending outside. The interior of the motor has an enclosed cooling system with an

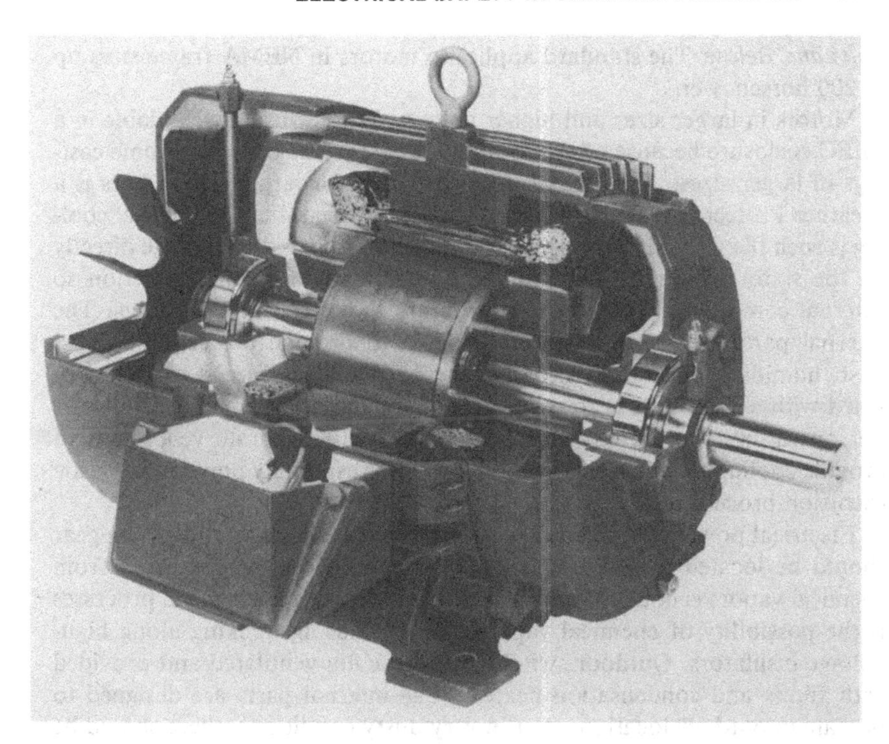

FIGURE 6-5. Cross-section of a chemical service motor designed for the severe service of a chemical process. The motor is an induction motor rated to operate on a three-phase, 60-hertz, 460-volt power system. The motor enclosure is TEFC, which provides protection against the entrance of chemical vapors and dusts. The fan blade on the left end should be of a nonsparking material. The rotor in the center of the motor is made of steel laminations and cast aluminum conductors that extend to the fan blades and out the end of the rotor. The bearings on each end of the shaft are grease-lubricated ball bearings designed for long service. A lifting eye is on the top of the motor, and an oversized terminal box is on the side. (Reliance Electric Company, Cleveland, Ohio.)

internal fan. One end of the shaft has the fan, which is guarded and blows cooling air across the length of the motor. The motor enclosure is not totally vapor right but does provide excellent protection against the outside environment. Many of the parts, including the frame, are cast iron.

Chemical service means that the external parts are constructed of materials that survive well in a chemical environment. The chemical industry has developed a standard for specifying motors for chemical service. This standard is ANSI/IEEE Standard 841-1986, *IEEE Recommended Practice for Chemical Industry Severe Duty Squirrel Cage Induction Motors 600*

Volts and Below. The standard applies to motors in NEMA frame sizes up to 200 horsepower.

Motors in larger sizes and higher voltages are not usually available in a TEFC enclosure because of the prohibitive cost of the cast-iron frame castings in larger sizes. The type of enclosure used for larger size motors is a Weather Protected II (WPII). This enclosure is open, but the internal cooling is such that outside cooling air is directed so it does not impinge directly on the stator windings. The internal air circulation changes direction so external contaminants will drop out before they reach the windings. The internal parts are treated and constructed to provide protection against dust, humidity, and other corrosive agents. The enclosure is usually provided with a filter and heaters to prevent condensation when the motor is not operating. Motors can also be provided with separate ventilation to provide cooling from a clean air source. This is especially popular in plastic extrusion processes.

Electrical power equipment like substation transformers and switch gear should be located at some distance from the process so corrosion from chemical vapors is not a major problem. One concern in chemical processes is the possibility of chemical vapors or dust causing tracking along high-voltage insulators. Outdoor switch gear is usually ventilated and provided with filters and condensation heaters. The internal parts are designed to operate in outdoor locations, but in very dusty conditions, there should be increased emphasis on cleaning internal parts.

Cable Trays and Cable Systems

Cable trays are manufactured of aluminum, galvanized steel, Polyvinyl-chloride PVC-coated metal, aluminum, steel, or fiberglass sections. Aluminum trays are popular because of their light weight and good conductivity. There are, however, some processes where aluminum can be attacked by chemicals, in particular, acetic acid, ammonium chloride, and ammonium hydroxide. Galvanized steel cable trays also can be attacked by certain chemicals. Fiberglass on PVC-coated metal cable trays are especially applicable to cooling towers and similar highly corrosive situations. The support strength of fiberglass trays should be verified, and equipment grounding conductors need to be added to fiberglass cable trays.

Conduit can be galvanized rigid or intermediate steel, aluminum, plastic coated, or plastic. Aluminum and galvanized steel conduit are used throughout the chemical industry. Metal conduit is also available with internal and external plastic coating. The *NEC* terminology for various types of conduit is as follows.

Intermediate metal conduit—Article 345
Rigid metal conduit (aluminum or steel)—Article 346
Rigid nonmetallic conduit (plastic)—Article 347

Aluminum conduit is usually not used less than three-fourth inch in diameter. Rigid nonmetallic conduit requires that a ground conductor be used in the conduit. Plastic-coated conduit offers the advantage of the mechanical strength of metallic conduit but is provided with a plastic coating that provides corrosion protection. There is, however, a precaution in using this type of conduit in corrosive locations. This plastic coating on the conduit threads must always be removed in order to provide metal-to-metal contact when sections of conduit are connected together. If conduit sections are not bonded together, the safety grounding system will be compromised and violate the *NEC*.

Modern cable systems use plastic insulations that have excellent durability in a variety of chemical environments, but there are chemicals that can attack the insulations on wire and cable. In one situation, a plasticizer attacked the PVC jacket on multiconductor tray cable and caused the jacket to deteriorate. The cable system had to be replaced or protected from the plasticizer. In other situation, hot heat transfer fluid sprayed from a failed pump seal and melted the insulation from a tray cable. The bare conductors touched, producing a spark that caused a fire. It is absolutely essential to know that the process fluids are compatible with electrical insulations.

Environmental Protection of Electrical and Process Control Rooms

Electrical and process control rooms are centralized locations for electrical and process control equipment, including process computer and control equipment, operator's consoles, and motor control centers. The equipment in these rooms is provided in enclosures with minimal environmental protection. It is usually guarded only for personnel protection and containment of electrical faults. Electrical enclosures may be gasketed.

If outside air containing chemical vapors, dusts, or humidity in sufficient quantities is allowed to enter the room, the reliability of the electrical and electronic equipment can be affected. Some chemical fumes can attack the contacts and connections on computer-printed circuit boards; some can attack electrical wire insulation and eventually cause failures. In some cases, the fumes may affect the health of personnel in the rooms. Humidity or rain can cause wiring failures and possibly shutdowns or fires. Protection against chemical fumes can be provided by locating the rooms or building some distance from the process unit; 50 feet is considered a minimum distance.

Most electrical and process control rooms in process areas are pressurized by ventilation and air-conditioning systems. Air conditioning is needed for the comfort of personnel and to cool electronic equipment. The air pickup point for these systems should be at a location where it is not exposed to potential vapor, gas, or dust releases. In highly congested process facilities, it may be necessary to provide tall stacks to ensure that the intake, makeup air is safe. Gas analyzers and special vapor filters can also be used to ensure that the inlet air does not contain harmful vapors.

The positive pressure inside the rooms should be at least 0.1 inch of water, but it should not be so high that it prevents operators from opening doors. The upper limit of pressure is 0.3 or 0.4 inches of water. Refer to NFPA 496 for additional information concerning room pressurization.

Sealing electrical rooms is essential to provide protection from the outside. A difficult area to seal is where cables or conduit enter the room, since there is usually a high density of cabling that enters an electrical or process control room.

Sealing around cables, especially large cable tray installations, requires special materials and care in the installation. Silicone seals that are applied like caulking can be used. UL listed foams can also be used.

A device called through-wall-barrier, which consists of a metal clamp that forces an elastomer and a seal around the cable, can be used. These materials are usually fire rated so they can be used to maintain the fire rating of a wall where conduits or cables pierce a firewall. Regardless of the method, it is essential to consider both the ease of installation and the addition of cables after the initial installation is complete. Seal bags are available, which provide ease of installation and can be removed and repacked to add new cables.

Process vapors can sometimes find devious ways to enter a control room. One such situation occurred when a process thermowell failed; the process fluid (which was toxic) pressurized the thermocouple head and forced the fluid through the cable by wick action, through the insulation packing between the conductors, and finally out into the control room.

Leakage of process fluid when the primary element has failed has occurred on a number of occasions. Conduit can act as a pipe for these fluids. Explosion seals leak at a small rate when exposed to a constant process pressure. It may take hours for significant leakage to occur, but it has happened. Never provide a direct connection from a process sensor into a control or electrical room. *NEC* Article 501-5(f) (3) indicates that a drain or vent is needed so the process seal leakage is obvious; then a second cable seal is required.

Process vapors or liquids have also entered control rooms through underground lines. In one incident, flammable vapors entered through a floor drain. In another, they entered through toilet connections. The backup of vapors occurred during a severe rain storm. In both cases, the control rooms were destroyed.

Process piping or other piping should not be allowed in electrical or process control rooms. The possible leakage of process fluids could cause fires or necessitate evacuation of the rooms. Article 384-2 of the NEC does not permit piping or ducts to be installed in rooms with switchboards or panel boards. The intention of this article is to protect electrical equipment from leaking pipelines. Although the article only applies to switchboards and panel boards, the principle applies to any electrical equipment. (An obvious exception is a sprinkler system intended to provide fire protection in electrical rooms.)

Electrical System Protection

Short circuits can occur because of electrical insulation failures, misconnections, or equipment failures. The *NEC* requires electrical systems to be protected against the sustained arcing and burning that can occur as a result of short circuits. The protection consists of fuses, circuit breakers, ground fault protection, and electrical system and equipment grounding. This protective system will minimize property damage, injury to people, and interruption of production.

The results of short circuits and insulation failures can be catastrophic. One chemical company reported $2.2 million in business interruption losses due to electrical power system failures in 1 year. In one situation, an operator was locking out a motor by opening the motor power disconnect. As he operated the disconnect, a wire broke inside the unit and shorted from a power line terminal to the metal enclosure. This caused arcing that continued and resulted in a burnout of the motor control center (MCC) bus bar system. Three inches of bus bar burned away. An investigation indicated that the MCC was not grounded. Grounding connections were added.

In another situation, an operator was shocked when he touched an electrical panel. An investigation indicated that water had entered the panel and bridged from a hot wire terminal to the metal enclosure. The enclosure was not grounded; therefore, the power supply breakers did not open to deenergize the panel. The panel was enclosed to prevent water from entering, and grounding connections were added.

Please note that these are not classified locations. Flammable gases, vapors, or dusts did not start the fires. It is important to understand that arcing and heating at a fault can initiate a fire in the combustible material within the enclosure.

Electrical cables, wiring, windings, and components are insulated from their metal enclosures. If for any reason the insulation fails, energized conductors, contacts, or terminals can find a conductive path to the metal enclosure. This is called a ground fault. Energized conductors or terminals can also find conductive paths to other energized conductors that are at a different phase voltage. In power system terminology, this is a line-to-line fault. Most power system insulation failures start at ground faults, but they can develop into line-to-line faults because the initial ground fault arcing damages the line-to-line insulation system.

If the conductors and equipment that are part of the faulted circuit are not deenergized quickly, sustained heating, arcing, and mechanical damage can occur. The amount of damage is directly proportional to the current squared multiplied by the time duration. Tremendous heating and mechanical forces are involved. The equipment may be damaged beyond repair; total burndowns and fires are possible.

These types of failures can occur in a residence, commercial establishment, or industrial facility, but they are especially damaging to chemical processes because of the possible adverse safety and environmental effects of the loss of power and electrical equipment fires. Transformer fires can release fluids and products of combustion that are environmentally unacceptable, and cable fires can release toxic vapors. The outage may be short term or long term, perhaps weeks, if major equipment is damaged.

Chemical processes also impose severe environmental and operating conditions on the electrical system. Chemical corrosion, hosedown, and dusts encourage insulation failures and can corrode grounding connections.

Many chemical processes operate continuously, 24 hours a day, 7 days a week. Therefore, maintenance of the electrical system may require working on the energized equipment, an undesirable alternative, or providing alternate power sources so the equipment to be maintained can be deenergized. Lack of maintenance is a factor in insulation and equipment failures. For these reasons, chemical processes demand special attention to electrical protection.

Electrical power equipment like substation transformers, high-voltage switch gear, motor control centers, and power distribution equipment operate at voltage and current levels capable of high-energy sparking and arcing. An electrical arc has the highest temperature on earth next to a nuclear reaction. A sustained short circuit can produce arcing that can melt and vaporize the metal conductors. Steel cabinets containing electrical equipment can explode. Electric power arcs can generate such high temperatures that the radiated energy can produce fatal burns to people up to 5 feet from the arc and major burns to people up to 10 feet away.

Preventions of short circuits caused primarily by insulation failures is the first line of defense. This requires good design, installation, maintenance, and work practices. Compliance with codes and standards and established safe practices are essential. Some typical types of insulation failures and possible causes follow:

Failure Moisture or water in enclosure bridges between live terminals and the metal case of the enclosure.

Cause Enclosure was not watertight (NEMA 3 or 4), or gasketing failed due to corrosion or damage, or the cover was left off or open. (Note: Enclosures with latched covers and hinged doors have a lesser chance of this happening.)

Failure Insulation failure due to vapors or liquids.

Cause Failure to evaluate chemical susceptibility of insulation.

Failure Insulation damaged during installation.

Cause Poor workmanship during installation, or inadequate working space or wiring congestion. (Note: This type of failure should have been caught during inspections or cable testing.)

Failure Insulation failure due to overheating.

Cause Overloading of wiring or equipment, or excessive ambient temperature, or poor cooling, loss of ventilation.

The ultimate life of electrical insulation is determined by its operating temperature. Cable, wiring, windings, and so on, are designed to operate with certain temperatures based on a maximum ambient air temperature of 40°C, adequate cooling air, and operation of rated load. Operation at higher temperature reduces insulation life.

Failure Loose connections.

Cause Poor workmanship. Failure to follow proper torquing of connections.

Failure Overvoltages caused by lightning discharges or power system switching can cause insulation breakdown.

Cause Failure to provide surge capacitors or arresters or inadequate grounding.

Failure Overcrowded wiring can cause conductors to break and come loose from their connections and short to ground or other connections.

Cause Failure to follow *NEC* rules for wire density.

Failure Tools slip—slip of a screwdriber or test leads can cause a fault to occur in electrical equipment.

Cause Failure to use proper tools; established hot work practices. Wiring or equipment density and inadequate working space could also be factors.

Regardless of the cause, insulation systems fail, and the circuit and equipment protection consisting of circuit breakers or fuses and proper grounding must deenergize the faulted circuits to minimize damage and the duration of the outage and to protect personnel. This protection is required by the *NEC*.

Circuit and Equipment Protection

Circuit and equipment protection, including the application of fuses, circuit breakers, and grounding, has been studied within the electrical industry, especially within the IEEE, for a number of years. The technology of circuit protection was greatly enhanced by the work of Kaufmann and other pioneers in electrical protection. The studies still continue; new equipment using microprocessor technology is now available that provides new options.

The *NEC* provides the basic criteria for the design and installation of circuits and equipment; UL lists fuses, circuit breakers, and other protective equipment; IEEE provides numerous standards and technical papers on the how to of circuit and equipment protection. Compliance with these standards, including UL listing of equipment, is essential; however, each user must apply these criteria with good engineering judgment.

Sound maintenance and updating the design as new equipment is added are the individual user's responsibility. Electrical systems do not get respect in some chemical-process facilities. The power is on; therefore, everything must be okay (until a transformer or switch gear blows up). An exciting development in this area is the number of personal computer programs available to engineer and document circuit and equipment design. These programs can develop coordination curves and other valuable data and can indicate problems in the electrical system; again, they must be applied with good engineering judgment.

The following is a discussion of the fundamental principles and terminology of electrical circuit and equipment protection. To begin, here are three goals of electrical circuit and equipment protection:

1. Provide devices (fuses and circuit breakers) in association with the grounding and ground fault protection equipment that protect the electrical system against permanent damage resulting from a short circuit, minimize arcing and burning, and protect personnel.

2. Provide fuses and circuit breakers rated to withstand the very high currents (possibly 50 times normal current) associated with short circuits without damage or blowing up (i.e., they have adequate interrupting capacity).
3. Provide a coordinated system design so current and time ratings of breakers and fuses are such that only the faulted circuit is deenergized and other circuits remain energized. Note that the current settings or ratings may be in thousands of amperes, and time intervals may be less than one-half cycle or 8.25 milliseconds.

Properly designed, installed, and maintained systems complying with the *NEC,* IEEE, and UL standards do not blow up or burn down when a short circuit or insulation failure occurs. If a motor power cable fails, only the motor is shut down, not the entire substation or plant. If a fuse or breakers open due to a short circuit, the motor does not blow up. Any cables, switches, or other equipment involved in the short circuit are not permanently damaged. The short circuit current magnitude and duration must be less than the withstand ratings of the equipment and cabling in the faulted circuit.

Prevention of short circuits is the first line of defense, but some failures always occur. Therefore, electrical protection is required.

The *NEC* provides the basic criteria for protection. Article 110-9 indicates that devices intended to break current shall have the interrupting capacity sufficient for the current and voltage to be interrupted. The current to be interrupted is the short circuit current at that point in the system. A short circuit study and analysis involving extensive knowledge of the electrical circuit parameters under fault conditions are required. The magnitude of the short circuit current is a function of the utility capacity energy contributions from any motors connected to the system, transformer capacity in kilovolt ampere and impedance, and cable impedances. A power system diagram called an impedance diagram is used to help perform laborious calculations.

Fuses and circuit breakers must have interrupting ratings that exceed the calculated short circuit current. The larger the transformer that energizes the system, the greater the short circuit current. For instance, a 1000 kilovolt ampere transformer would require an interrupting capacity of approximately 33 kiloamps at the transformer secondary. A 2000 kilovolt ampere transformer would require twice the interrupting capacity. For this reason, it is prudent to use multiple units instead of a single large unit.

The *NEC* also indicates that overcurrent protective devices, component short circuit withstand ratings, electrical circuits total impedance, and other circuit characteristics must be selected and coordinated so the circuit protective devices clear a fault without extensive damage to the components of the circuit.

The short circuit or fault can occur between two or more conductors or between circuit conductors and the metal enclosure or raceway. The short-circuit withstand rating for conductors, switches, circuit breakers, and so on, is the maximum short circuit current the equipment can tolerate without sustaining permanent damage. For cables and wires, the limitation is temperature rise; for equipment, it is mechanical forces and temperature. The withstand rating is given in current and time intervals.

The *NEC* has a number of requirements for the protection of cables, transformers, and other equipment. Section 240-12, Electrical System Coordination, indicates that if an orderly shut down is required to minimize hazards to equipment and people, a system coordination based on coordinated short circuit protection and overload indication shall be permitted.

Coordination localizes a fault condition to restrict outages to the faulted equipment and is accomplished by engineering the selection and setting of protective breakers and fuses so the fault currents only open the breaker or fuse energizing the faulted circuit.

Selective coordination is essential in any chemical-process facility. Serious safety and operational problems are possible if a failure of one equipment item causes other parts of the process or the entire process to shut down unnecessarily. Figure 6–6 shows a nonselective system.

Selective coordination requires that the upstream protective devices be time delayed in their operation to permit breakers closer to the loads to operate quickly. To minimize damage in the event of a short circuit, however, fast operation is desirable. Therefore, selective coordination compromises protection, and experience and judgment are essential in establishing settings and equipment selection. Figure 6–7 shows a selectively coordinated power system.

Fuses and circuit breakers are used to protect cables, power panels, motors, and transfer switches, and each has its own characteristic and area of application. A fuse is a protective device that interrupts current by fusing open its current-sensitive element when an overcurrent or short circuit occurs. When the fuse element opens, it must be replaced. Fuses only open the phase or circuit to which they are wired. They are direct acting and have no external devices. They respond to current and time and have current–time curves determined by the design of the internal fusible link. They sense and interrupt the current in one self-contained device; external connections are not required. The fuse link that opens the circuit is contained within the body of the fuse. Therefore, there are no open contacts that can corrode or ignite flammable vapors or dusts.

There are a number of different types of fuses for various types of loads

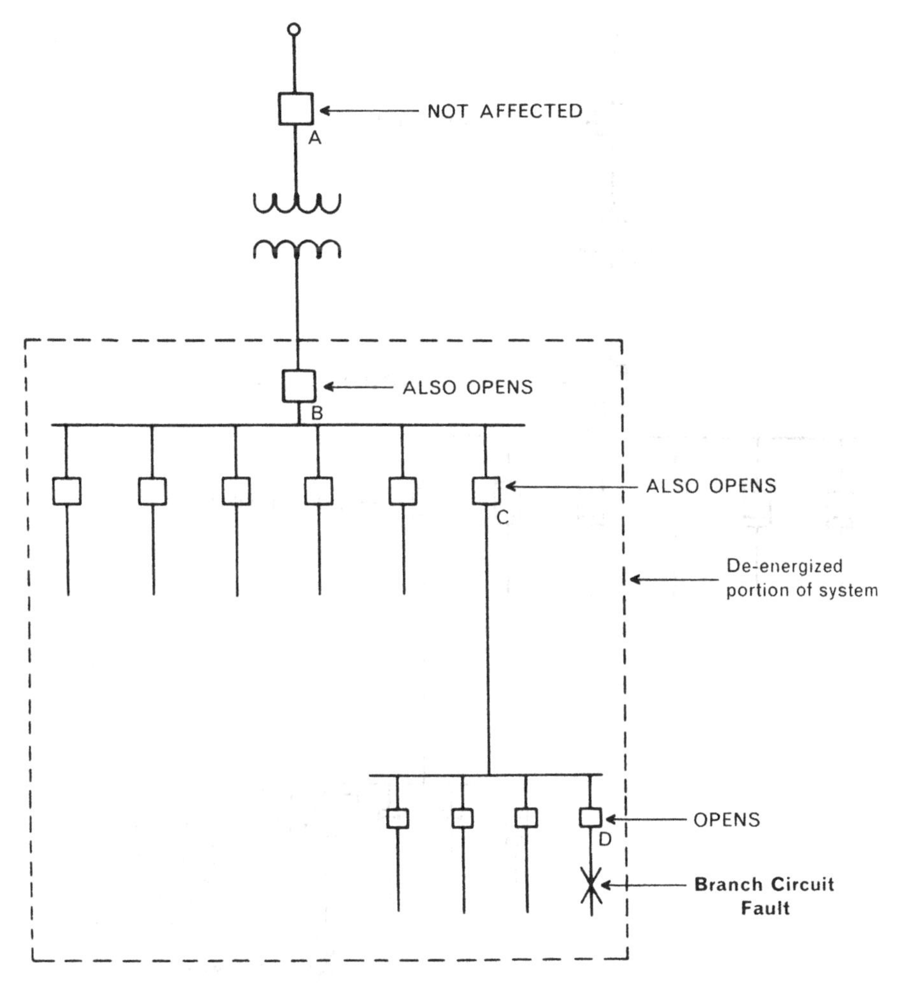

FIGURE 6-6. Nonselective electrical power system, which should not happen in a chemical-process facility. (Bussman Division, Cooper Industries, Inc., Selective Coordination of Overcurrent Protective Devices for Low Voltage Systems, Bulletin EDP-2, 5.)

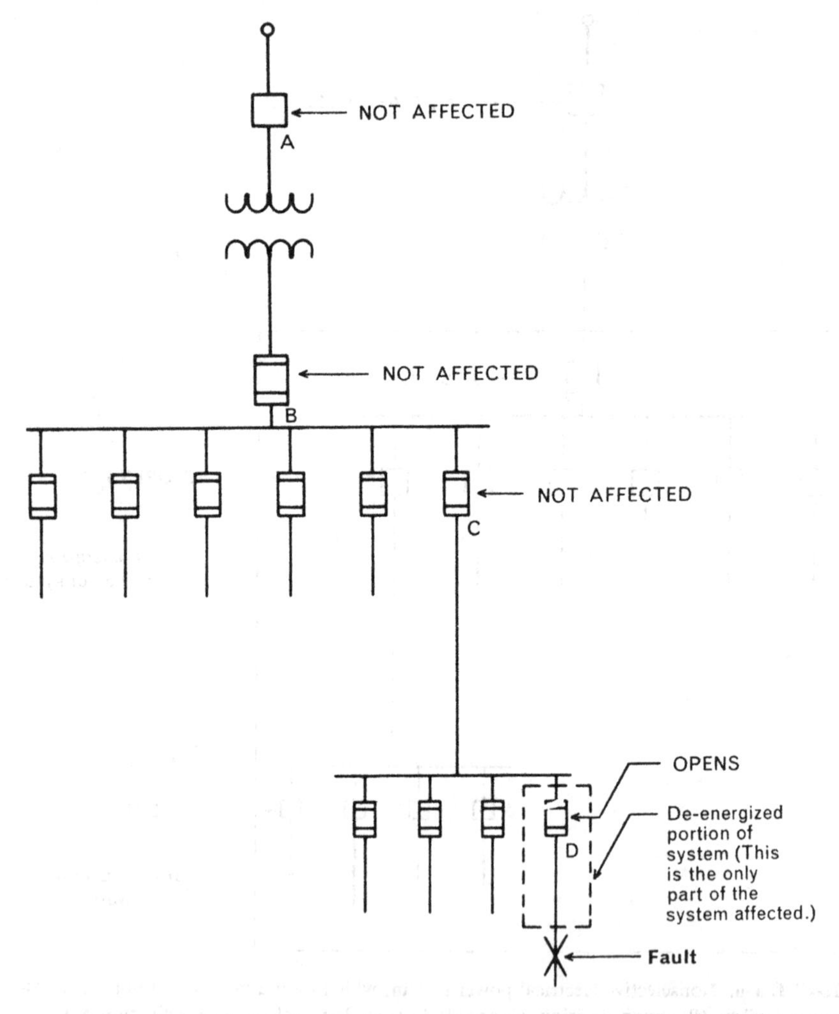

FIGURE 6-7. Properly designed, selectively coordinated power system for chemical-process facilities. (Bussmann Division, Cooper Industries, Inc., Bulletin EDP-2, 7.)

as covered by UL/ANSI standards. Fuses can provide overload protection for motors and transformers, but they are especially adapted to short circuit protection, especially where high interrupting capacity and speed of operation are required.

Current-limiting fuses are especially effective in limiting the energy in a fault. The design of the fusible link is such that it interrupts the fault current with one-half cycle before if can reach its peak value. These fuses significantly reduce the energy to the fault to a fraction of what it might be if noncurrent-limiting fuses or circuit breakers were used. The term i^2t is used to represent the actual thermal stresses on the faulted equipment during the first few cycles of a fault. Fuse curves and data also provide the means to determine the let-through current as a function of available fault current.

Fuses require replacement with the same type and size. They usually do not provide any indication they have operated. Therefore, testing is usually required to determine that a fuse opened. Fuse operation can also cause a condition of single phasing (one phase open), which can, under certain circumstances, cause equipment misoperation and failure.

Circuit breakers can be used to protect electrical systems against the damaging effects of short circuits. They sense the current and open movable contacts. Also, like fuses, circuit breakers are covered by the NEC, UL, and IEEE standards; UL-listed equipment should always be specified. Circuit breaker contacts can be opened and closed manually and function as a disconnect, giving circuit breakers an advantage over fuses. Circuit breakers have interrupting ratings, which must be greater than the available fault current. They also have a momentary rating, which indicates the ability to withstand mechanical stresses during the first few cycles of a short circuit. Circuit breakers, like fuses, are rated in voltage, continuous current, and interrupting amperes.

There are three types of low-voltage circuit breakers: molded-case, insulated-case, and power circuit breakers. Molded-case circuit breakers are integral units in plastic housings. They are used to protect individual loads like distributing panels, transformers, and motors. These circuit breakers can be tripped by a current-sensitive, bimetallic trip or a solenoid operated trip. Solid-state trips, available in many sizes, can have their time–current characteristics easily adjusted to meet various selective coordination schemes.

Circuit breakers are generally slower acting and have less interrupting capacity than fuses. In some situations, current-limiting fuses are required to protect breakers. Current-limiting, high-interrupting capacity breakers are, however, available.

The main advantage of circuit breakers is the ability to adjust the time–current response of the breaker to provide selective coordination with downstream equipment. Using solid-state electronics, a circuit breaker response to short circuits can be adjusted to meet the most difficult coordination requirements. The newer solid-state systems provide greater flexibility than older electromechanical systems. Solid-state trip units can be set with current–time settings to develop a time–current curve to match difficult applications.

Another significant advantage of circuit breakers is their adaptability to provide a ground fault protection scheme.

The structural features of low-voltage and medium-voltage switch gear must provide isolation, separation, and guarding of the energized components. The power circuit wiring is usually insulated and the bus should be enclosed and have grounded metal barriers. All live parts should be enclosed in metal compartments that are grounded. Automatic shutters to prevent exposure to power circuits should be provided wherever possible. Mechanical interlocks should be provided to ensure a proper and safe operating sequence. Meters and other control devices and their wiring should be isolated by grounded metal barriers. Metal-clad switch gear describes power equipment that includes these safety features.

Low-voltage and high-voltage circuit breakers are typically located in substations to provide primary protection and secondary distribution to MCC and other large loads.

Electrical Grounding

Grounding is an important safety issue, particularly in chemical process facilities. Incorrect grounding, particularly no grounding, has serious consequences. But electrical grounding is a controversial topic. There have been and will continue to be numerous discussions and questions among engineers, electricians, electrical inspectors, and other people concerned with electrical business, on grounding. Article 250 of the *NEC* is dedicated to grounding, and many other articles in the *NEC* have grounding sections. There are numerous standards, articles, technical papers, and books on the subject issued by the IEEE, ISA, other technical associations, and trade publishers. In spite of this effort, it remains a controversial and often misunderstood topic.

Grounding covers a wide spectrum of applications. It is involved in lightning protection, static spark prevention, power system protection, and electrical noise reduction. Each of these topics involve different problems.

Lightning protection grounding of buildings and structures consists of providing contact between earth and the lightning rod and down conductor system to direct the lightning stroke to earth. Static electricity grounding consists of interconnecting, or bonding, the process piping, equipment, and structure together, then to earth to minimize static charge accumulation and prevent static sparks that could ignite flammable gases, vapors, or dusts. The above grounding and bonding together of metallic objects does not directly involve electrical equipment or wiring. In fact, it might be referred to as bonding and earthing instead of grounding.

Grounding of electrical equipment involves the interconnection of electrical equipment, enclosures, cabling systems, conduit, cable tray, and so on, together and to earth or to a conducting body that serves in place of earth, typically the building structural steel. Electrical equipment grounding protects against excessive arcing, sparking, burndowns, and shock voltages that can occur when a wiring or equipment ground fault occurs.

Electrical noise prevention grounding involves the interconnection of electronic instrumentation and process computer systems, including enclosures and wiring systems, together and to earth or to a conducting body that serves in place of earth to minimize high-frequency voltages that can disrupt and interfere with the signal wiring and possibly damage components. Electrical equipment grounding and noise grounding are not independent issues. Poor power grounding can cause noise and interference in signal wiring.

Proper terminology is important in any subject, but it is especially so with grounding. The *NEC* defines ground as a conducting connection either accidental or intentional between an electrical circuit or equipment and earth or to a conducting body that serves in place of earth. Bonding is the conductive connection of metallic objects so as to minimize voltage differences and conduct any current imposed on the bonding connection and conductors adequately without sparking, arcing, overheating, or blowing apart. Low resistance is not as important in bonding as the reliability and durability of the bond. Bonding jumpers and connections are exposed to severe environmental conditions in chemical processes. The size (cross-sectional area) of bonding jumpers is primarily based on durability. They are oversized relative to the current they must conduct. Regardless of the type of grounding, bonding, or earthing, the ultimate goal is to prevent harmful voltage differences and to conduct fault currents without arcing, sparking, excessive temperatures, or burning.

All grounding involves the interconnection of metallic objects. Plastic or other nonconductive objects cannot be bonded or grounded.

Electrical System Grounding

Electrical system grounding is covered in Article 250 of the *NEC* and in IEEE standards. Article 250 is concerned with the protection of people and property from the adverse effects of ground faults. Electrical power and control system wiring and circuits are insulated from metallic enclosures and wireways, building structural steel, and earth. When for various reasons the insulation system fails, large short circuit currents flow through the ground system back to the power source. Large ground fault currents can cause arcing and sparking and create strong magnetic forces between conductors and shock voltage differences between enclosures.

Prevention of ground faults by properly designing the wiring system and selecting the enclosure and by providing safe working clearance and good maintenance practices can do much to prevent failures; however, some ground faults will still occur. Therefore, grounding systems are essential and are required by the *NEC.*

Designing a safe system requires a coordinated design taking into consideration the capacity of the system (usually expressed in short circuit millivolt amperes), characteristics of the breakers and fuses, and the impedance of the ground path, including the phase conductors and equipment ground that. The first and most important distinction is that of system grounding versus electrical equipment grounding.

System Grounding

System grounding is the connection of the insulated wiring system to earth or to a building structure or other conducting body that serves in the place of earth. The phase conductor connected to ground is called the grounded conductor. Typically, it is the transformer neutral; corner delta grounding has been used by some companies. For identification purposes, the *NEC* requires that the wire insulation be white, gray, or colored.

System grounding stabilizes the voltage of the wiring system to earth. A transformer secondary is a wiring system; in *NEC* terminology, it is called a separately derived system. System grounding protects the wiring system against external disturbances that could cause an overvoltage on the system relative to earth. For instance, if a lightning strike were to cause an overvoltage to earth on the transformer secondary, the system ground connection would channel the overvoltage to earth. If the high-voltage primary wiring in the transformer were to fail and contact the transformer secondary low-voltage, the system ground would prevent the secondary voltage from reaching a high value relative to earth.

Electrical system grounding can be

Solidly grounded
Ungrounded
High-resistance grounded

Other specialized types of systems are mentioned in the IEEE grounding standards, but the above types of systems describe the fundamental principles involved.

Solidly grounded systems are directly connected from the transformer neutral connection to earth or building steel connected to earth. They provide voltage stabilization since the circuit is tied directly to earth. A ground fault acting through the equipment grounding system will open the circuit protection and quickly clear the fault and shut down the system (Figure 6-8).

An alternate to the solidly grounded system is the ungrounded system. Some older facilities still use this type of system grounding. The advantage of this system is if a ground fault occurs, large ground currents do not flow, the circuit protection does not operate, and the system does not shut down immediately. A planned shutdown is, however, required to repair the ground fault. This first ground fault is detected and alarmed by a standard arrangement of indicating lights and alarm relays. A scheduled shutdown can then take place, at which time the ground fault can be found and corrected. It is essential that maintenance personnel be trained in finding ground faults in a reasonable time. This is done by using specialized instruments developed for this purpose.

There are risks associated with an ungrounded system. If a second phase fails to ground before the first phase fault is corrected, a line-to-line short occurs, producing higher fault currents and a total shutdown. The damage is much greater than a line-to-ground fault.

Voltage stability is another risk. No system is truly ungrounded. The wiring systems are capacitively coupled to wireways, earth, and building

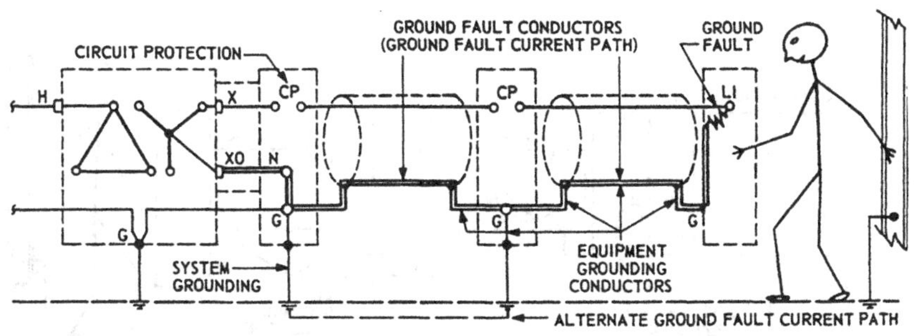

FIGURE 6-8. Fault current path for a solidly grounded system. Note the ground fault current follows the path of the equipment grounding conductors. (West, R. B., Equipment Grounding for Reliable Ground-Fault Protection in Electrical Systems Below 600 V. *IEEE Transactions on Industry Applications* IA-10, no. 2, Mar./Apr. 1974, p. 177, Fig. 2.) © 1974 IEEE.

steel. It is always possible to measure a voltage to ground on ungrounded circuits. If the distributed capacitance of each phase were equal, each phase-to-ground voltage would be 277 volts on a 480-volt system, as it would be on a solidly grounded system. Note that after the first ground fault, the other phases are at full line voltage to ground. An ungrounded floating wiring system can produce another serious problem. If the first ground fault is an arcing fault instead of a solid connection, it will react with the system distributed capacitance to produce serious overvoltages. In one plant, an arcing fault on an ungrounded system damaged 58 motors.

The high-resistance grounded system offers the best features of the ungrounded and solidly grounded systems. With this system, a grounding resistor is inserted between the transformer neutral and ground to limit the ground fault current to a low value, typically less than 5 amperes. This will not trip the circuit protection (see Figure 6-9). The connection to ground stabilizes the voltage, but the first ground fault does not interrupt service. The ground fault must be detected and corrected before a second phase goes to ground. Therefore, a ground current detection relay and alarm should be provided.

Electrical Equipment Grounding

Electrical equipment grounding is distinct from system grounding. Equipment grounding is the interconnecting of all metal parts enclosing electrical conductors, components, or equipment into a series network that provides a ground path back to the supply transformer for any ground fault current that could occur. It must be able to conduct any ground fault current that might occur without sparking or arcing and must have a low enough imped-

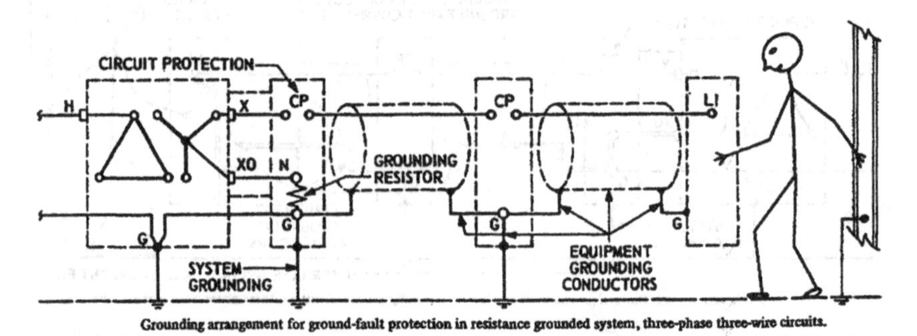

FIGURE 6-9. Resistance grounded system. The grounding resistor limits the ground fault current. This system provides the voltage stability of a solidly grounded system and the continuity of service of an ungrounded system. (178, Fig. 5.)

ance to prevent shock voltages and facilitate fast operation of circuit protection, fuses, or breakers.

The *NEC* Article 250 describes the types of systems required to be grounded; equipment grounding is, required for all equipment or components that operate at more than 50 volts ac whether the system is grounded or not. The *NEC* also describes the types of acceptable equipment grounding conductors, including cable trays, conduit, flexible conduit, and individual conductors. Individual conductors can be bare conductor or insulated. If insulated, the insulation color must be green or green with a yellow stripe. The equipment grounding system must follow these stipulations:

1. It must be continuous back to the supply transformer. All metallic parts must be bonded together. Where there may be gaps, bonding jumpers must be provided.
2. All parts of the system, especially any connections, must be corrosion resistant to the particular environment.
3. All connections must be wrench tight and not allow sparking or arcing during a ground fault. The impedance of the system must be low enough to permit the fuse or breaker to operate quickly and to prevent shock voltages between enclosures.

The impedance of the ground circuit is influenced by the spacing between the faulted phase conductor and the equipment ground conductor. The effect increases with the current in the conductors. For instance, the impedance triples for a 200-ampere cable if the spacing to the ground conductor is increased from 1 inch to 8 feet. This is why the *NEC* requires that the equipment grounding conductors and phase conductors be run together in the same conduit or cable tray. Metallic cable trays are excellent equipment grounding conductors, especially aluminum trays, due to their large cross-sectional area. Cable trays have low impedance and can withstand significant fault currents if they are properly bonded. Cable trays used as equipment grounding conductors must be marked as such.

The *NEC* requires that the neutral or grounded conductor be grounded at one, and only one, point for separately derived systems. The only exception is at the utility service to the plant. If the neutral is connected to ground at more than one point, it will cause stray currents to flow in the equipment grounding system and in the structure. Stray currents (referred to in the *NEC* as objectionable currents) through the equipment grounding conductors can cause electromagnetic interference with electronic instrumentation and computer systems, sparking and arcing in the process structure, damage to load cells or motor bearings, and other undesirable effects. The lower the impedance of the equipment grounding conductor, the lower the amount of stray current in these other paths during ground faults. Equipment grounding should only carry current during a ground fault, and most of the ground fault current should flow through the equipment ground conductors.

Earth is a poor conductor and does not provide a low-impedance path. For this reason, the *NEC* does not allow earth or the structural steel members of a building to be the only equipment grounding conductor. Thus, the practice of driving ground rods at motors or other equipment as a sole equipment ground conductor can cause dangerous voltages and is a violation of the *NEC*. Additional equipment grounding at the equipment is acceptable to ensure that voltage differences do not occur but the equipment grounding conductor must be in place. This supplemental equipment grounding is sometimes used for voltages of more than 600 volts.

Grounding Electrodes

The path from the grounding system to earth (system grounding) is by means of a grounding electrode that provides contact with earth. The grounding electrode can be the nearest grounded structural steel, a metal water pipe system, concrete-encased rods, a ground ring of conductor around the building or structure, electrodes consisting of ground rods, plate electrodes, or metal underground systems. If a single ground rod is used and it does not have a resistance to earth of 25 ohms or less, a second rod should be added.

Substation grounding practice, as described in IEEE standards, requires a very low resistance to earth, typically less than 1 ohm. Substation grounding usually consists of a ground mat system to achieve minimum voltage differences under very high fault current and voltages that could occur at a substation. Bonding all metal parts, fences, and so on, together to minimize voltage differences that could occur in the event of a power fault is essential.

All ground electrode systems consist of underground metal or structures attached to underground metal. Wherever possible, underground piping or other metal object should be bonded together to avoid voltage differences that could occur. The larger the electrode system, the lower the resistance to earth.

Underground electrode systems can also contribute to underground corrosion since buried electrodes become part of underground electrolytic cell action. Copper can be an especially significant contributor to underground corrosion. The concrete-encased rebar grounding electrode system, referred to as the Ufer ground, has significant advantages in terms of low resistance over other electrode systems, especially in highly corrosive soil conditions.

This system uses the rod rebar in building steel foundations connected together and to the building steel columns to form a grounding electrode system. The reason this system is effective is that concrete has the same or higher conductivity as earth, especially in highly corrosive soil conditions. The electrode system is particularly effective because it uses an extensive network of rebar and building steel. It is only necessary to connect the rebar and columns together. The addition of ground rods is unnecessary. Resistance measurements on some of these systems are very low, less than 0.5 ohm.

Ground Fault Relaying

Ground fault relaying is used in power systems to protect against ground faults, particularly arcing ground faults. The system senses ground fault current and opens the circuit breaker to energize the faulted circuit. Sensing the ground current is accomplished by a current transformer (CT) in the main transformer's neutral-to-ground connection on a three-phase, three-wire system or by enclosing the phase conductors and neutral in a donut current transformer to sense the differential current that would occur in a ground fault. The output current of these CT's trip ground relays which are adjusted to provide a coordinated ground fault protection scheme. The system design is intended to trip only the faulted circuit, but some systems are complicated and the possibility of tripping other circuits exists.

Ground fault relaying was developed primarily to protect against the destructive effects of arcing ground faults. Arcing ground faults have produced burn downs and significant damage. The arcing ground fault current usually is not large enough to trip the breaker or fuse. Therefore, the fault continues to burn without interruption. The ground fault current with an arcing fault is a fraction of the full line current, whereas a solid ground fault would be much larger than the line current.

Arcing ground fault burn downs have occurred on high-current (more than 1000 ampere) systems at bus bar systems in the main distribution equipment. For this reason; the *NEC* requires ground fault protection on feeders rated larger than 1000 amperes.

Static power conversion equipment, for instance, large dc drives, can be especially susceptible to arcing ground faults because of the number of devices connected to the high-current bus. Insulating bus bars can be effective in preventing bus bar insulation failures.

Equipment ground fault protection can also be provided on lower amperage, that is, 15–30 ampere breakers. This type of breaker trips at about 30 milliamps ground current and is used to protect electrical pipeline tracing, process heaters, or other loads against arcing faults and fires that can occur when the equipment grounding system is inadequate, arcing faults occur, or both conditions. This type of breaker was developed in response to fires that occurred with electrical pipeline tracing systems. These fires occurred when the electrical heat tracing outer insulation jackets failed and the exposed heater conductor split, opening the conductor and providing a sparking contact at the break. Wet piping insulation over the tape provides a conducting area over the split conductor. The resulting fires have been referred to as wet wire fires. The pipe or metallic braid on the tape does not provide an adequate equipment grounding path to trip a "normal" breaker. An equipment ground fault breaker that trips at a low current (30 milliamps) is required. This type of equipment ground fault protection is required by the *NEC* (as indicated in Article 427–22) on all electrical pipeline tracing that does not have a metallic jacket.

There have been burn downs on installations that have metallic jacketed pipe tracing tape. Therefore, equipment ground fault protection should also be provided for these installations. Equipment ground fault does not protect people from electrocution. The trip current (30 milliamps) is low enough to prevent arcing but high enough to prevent nuisance tripping due to leakage current across the cable connections. A GFCI *cannot* be used for this application because its trip current is too low. It would be tripped by the pipe heat tracing system leakage current.

Ground Fault Circuit Interruptors

Ground fault circuit interrupters are specialized circuit breakers that trip at a low current to prevent electrocution. They are referred to as "people protectors" and protect people against line-to-ground failures but not against line-to-line contact. GFCI receptacles are available at 120 volts, 15, 20, and 25 ampere sizes. They are especially effective in wet locations where there is a poor equipment grounding system (for instance, nonmetallic environments). They are required by the *NEC* in certain locations, especially wet locations. In chemical-process facilities, either a GFCI or an assured equipment grounding program is required to protect construction workers against shock due to wet conditions or electrical insulation failure principally with portable tools. OSHA requires GFCIs or an assured equipment grounding program on all 120-volt, single-phase, 15- and 20-ampere receptacle outlets on construction sites that are not a part of the permanent wiring of the building or structure. The assured equipment grounding program involves testing and visual inspection, including written procedures and documentation that the program has been implemented.

ELECTRICAL POWER RELIABILITY AND QUALITY

Electrical power system failures and abnormalities have caused serious safety problems in chemical-process facilities. Power interruptions to the plant and within the plant can shut down critical process equipment. Instrument and control system power supply abnormalities can cause misoperation and shutdown of critical safety controls.

The consequences of power disruptions depend on the type of process and its sensitivity to disturbances. In one situation, the loss of agitation at a critical time in an exothermic batch reactor cycle allowed a runaway reaction to occur, and the release of flammable materials resulted in an explosion and fire.

Some processes require a considerable length of time to stabilize, and even the slightest disruption can cause a line to produce below-grade product for hours. The loss of power, for example, resulted in plastic product solidifying in process piping and vessels. Equipment was disassembled and heated in burnout ovens to remove the solidified plastic.

Environmental releases can also occur during power disruptions. An example is a recent power failure that resulted in the shut down of a critical fan and the release of EPA-listed material.

Power failures and disruptions to control systems can also cause unsafe conditions. As an example, a battery system supplying control power to a gas turbine failed and, in combination with other circumstances, caused a destructive fire and a 12-week outage.

Power system reliability is also a personal safety issue. The loss of lighting in a control room or process area can pose serious safety risks to people.

In addition to safety implications, power disruptions are costly. A high-voltage line to a chemical plant failed when a power line contractor mistakenly dug into an underground substation control cable. At the same time, the alternate line to the plant was out of service. The entire plant shut down, resulting in a loss of over $1.5 million in property damage and $1 million in business interruption losses.

There are, however, protective measures to minimize the probability of power problems. In choosing among various alternatives, avoid overly complex solutions and always remember that protective systems can fail if they are poorly designed or not maintained and tested. An example is a situation where an UPS using a motor-generator (MG) set failed when a motor bearing failed. Every protective system must be capable of being tested and maintained.

Lighting Systems

During a power outage, there must be sufficient lighting in process areas, control rooms, electrical equipment rooms, and areas around substations to allow operators to exit process areas safely, to provide an orderly shut down of the process unit from the control room, and to permit maintenance and operation of main electrical distribution equipment. It is also important to ensure that process areas that are lighted by metal halide, mercury vapor, or other types of fixtures that require a restrike time after a power blip are provided with other types of lighting. Mercury vapor fixtures require approximately 20 minutes for the arc in the bulb to restrike; it is therefore advisable to provide a few incandescent or fluorescent fixtures, especially at stairs, for temporary ride-through lighting. High-pressure sodium fixtures can also be used since they have a short restrike time after a power dip. The *Life Safety Code,* NFPA 101, requires that the lights in a process area are powered by at least two circuits so if a circuit breaker or fuse opens, an entire process areas is not in darkness.

The *Life Safety Code* also requires emergency lighting for egress and evacuations. Typically, the areas include aisles, passageways, and stairs that lead to an exit. The lighting intensity is to be a minimum of 1 foot candle for a minimum of 1 ½ hours. Battery-operated lights using rechargeable batteries are typically used, but emergency power generators or alternate power sources are also permitted by the *NEC* under certain circumstances.

Battery-operated lights are available for classified locations. Fluorescent lights with battery-operated inverter modules are also available for control rooms, electrical rooms, and other indoor locations. They are especially adapted to low-ceiling applications. Some facilities use a centralized battery, battery charger, and controls instead of individual battery units. The centralized system may provide easier testing and maintenance. If an engine generator provides power for emergency lighting, the *NEC* requires the maximum delay for the engine generator to reach full speed and provide emergency lighting to be less than 10 seconds. Exit signs must also be illuminated or easily identified.

NEC Requirements for Emergency Systems

The electrical requirements for emergency systems are found in Article 700 of the *NEC.* This article does not require emergency lights or other emergency systems; however, where such lights are required by municipal, state, federal, or other codes or agencies, *NEC* Article 700 describes the types of wiring and maintenance practices to use. In addition to exit lights, emergency systems may include fire detection and alarm systems, communication systems, and other systems required for life safety in the event of an emergency. Article 700 requires the wiring for these systems be separate from other wiring and permits a separate service, an engine-generator set, or a UPS as an emergency supply in addition to batteries.

Article 701 of the *NEC* covers legally required standby systems that provide power for fire fighting, rescue, and other critical situations not considered emergency systems. Separate wiring systems are not required, and the maximum allowed time for the restoration of power is longer (60 seconds) than for emergency systems. Article 702 covers optional standby systems intended to provide alternate sources of power for industrial and commercial facilities where a power interruption could result in damage to the process or product.

The *NEC* does not address power system reliability or quality. Where they have an impact on chemical process safety and environmental releases, it has been and should be the responsibility of the engineers to identify the risks and needs. The reliability and quality of power to critical process equipment and control systems are a safety concern in the chemical industry.

Utility Power System Reliability

The failure of the utility power serving a chemical-process facility can have devastating results. This is not to say that power failure to an individual motor or power disturbance to a process computer is not serious, but plant power failures can cause the loss of instrument air, plant nitrogen supply, cooling tower water, steam, and other utilities. Utility power disruptions can vary from short blips to prolonged total outages. Electrical utility systems are large networks of power lines, towers, and transformers, covering many miles with a wide exposure to adverse conditions. Some outages are the result of weather conditions, lightning storms, wind, or ice, whereas others occur as the result of accidents or equipment failures. Weather accounts for the largest percentage of outages, followed by equipment failure and accidents. One utility reports the following outages at one plant location during 1 year:

Outage	Duration
Earthquake	30 seconds
Car hits pole	30 seconds
Lightning	30 seconds
Heavy wind storm	1 hour 45 minutes

During long periods of excessive power demand because of an extended heat wave, the electrical utility may find it necessary to practice brownouts (a 3–8% reduction in voltage) or periodic load shedding.

Utilities usually provide dual feeds to large processing plants and provide, by switching, alternate feeders and equipment. In one process facility, the utility can supply power from any of four different configurations. Knowledge of how the utility system is designed (a utility one-line diagram is essential), how the system is operated, what the emergency plans are, and when equipment is changed or maintenance is performed is important. Being aware of relaying and protection design and reclosing practice are also important. Reclosing practice refers to the opening and closing of breakers into a fault with the hope that the fault will clear itself between the opening and the closing. Reclosing practice needs to be coordinated with the large motor operation to prevent possible damage due to restarting of the motors. The utility outage record in terms of frequency and duration of outages is indicative of expected outages.

A relatively new concern of some electrical utilities is the power line harmonics that large adjustable speed motor drives or rectifier systems can inject into the power system. These harmonics can react with power factor capacitors and cause circuit resonance. The result of excessive harmonics can be misoperation of equipment, capacitor failures, and overheating of cables and equipment. Any time a large, adjustable-speed motor drive is added to a system, a harmonic study should be done. "Large" could be considered over 1000 horsepower or 20% of the system capacity. Any time a large motor is started in a process facility, it is also prudent to inform the utility.

Inplant Electrical Distribution Systems

Inplant electrical distribution systems consist of the main plant substation, which receives power from the utility system via two feeders at voltages between 4 and 260 kilovolts and reduces that voltage to a medium voltage typically between 4 and 14 kilovolts for distribution to individual process unit substations, which further reduce the voltage to 480 volts for process equipment motors and other loads. There are different medium voltage distribution system designs that provide varying degrees of backup and redundancy in the event of equipment failures and maintenance. The types include radial, primary selective, and secondary selective.

The radial system does not provide duplication of equipment; therefore, the process must be shut down to perform maintenance and servicing of equipment. The primary selective system provides redundant primary feeds. If the alternate primary feeder is energized from a different circuit breaker or feeder, it provides additional backup. The primary loop system provides alternate feed from either direction; the secondary selective system provides redundancy for the substation transformers and switch gear. Note that if the transformers are sized for the total load, either transformer can be isolated and removed for service. The main secondary circuit breaker provides isolation for the transformers. Switches or circuit breakers may be necessary to isolate equipment for maintenance. A good system design does not require a shut down for maintenance, nor should it be necessary to maintain equipment while it is energized.

Electrical System Reliability

The reliability of the inplant electrical power distribution system is a major factor in deciding how much backup is required. The reliability of the system depends on the age of the facilities, the degree of load (the percentage of full load rating), environmental conditions (dust, corrosion, rain, etc.), maintenance practices, quality of the original installation, and system protection. All electrical equipment has a life cycle that depends on the application and maintenance, but it is usually quite long, typically more than 20 years. If equipment like transformers are operated at reduced load, they can expect a longer life. Dust, corrosion, and rain can cause deterioration of enclosures and of tracking on high-voltage insulators. The quality of the original installation, especially the workmanship (e.g., of high-voltage cable terminations and stress cones), affect the reliability of the system. Proper maintenance is vital. Cleaning and servicing equipment, particularly transformer oil, is critical. Finally, system protection, grounding, circuit protection, and overvoltage surge protection are all essential.

Process Electrical System Design

The design of the low-voltage process electrical system should match the process. If multiple lines, reactors, and so on, are involved, then wherever possible and practical, individual and dedicated MCCs and control and power distribution equipment should be provided so that one failure does not shut down multiple units or lines.

This philosophy should extend to instrument power distribution. A simple fuse blowing should not shut down multiple lines or an entire unit. Wherever possible and practical, systems should be isolated. The system should also be designed with the disconnects and switches required to perform maintenance on equipment without shutting down an entire unit or forcing maintenance personnel to work on equipment while it is energized.

Power System Considerations for Large Motors

Electrical motors are by far the largest electrical load in a chemical-process facility. Motors drive pumps, fans, agitators, compressors, and so on, and essentially run the chemical processes.

When a motor starts, it draws five to six times its rated current until it reaches full speed. Motor starting time is determined by the inertia of the load. Pumps are usually easy loads to start. They reach full speed in a few seconds unless they are filled with polymer. Centrifuges and compressors with higher inertias may require 10–25 seconds to reach full speed. Large motors are a special concern in chemical processes, especially if the motor starts a high inertia load. "Large" is any motor that is more than 20% of the system capacity. For a typical 480-volt system, any motor in the range of 200–300 horsepower should be considered large. For larger voltage systems (2400 or 4160 volts), the size depends on the transformer kilovolt ampere.

Large motors produce significant voltage dips throughout the system. In addition, the longer the motor starting time, the greater the effect of the voltage dip. Any time the voltage dips below 90% of the nominal value, problems can occur. Relays and contactors drop out at typically 70–90% of nominal voltage. Computer or PLC systems can shut down or cause misoperation during voltage dips.

Repeated restarting of large motors can, under certain circumstances, cause damage to the motor as well as voltage dip problems. After a motor is deenergized, a residual voltage is present that decays with time. The decay period increases with the size of the motor but could be 1–3 seconds. If the voltage is reapplied because of a power dip or motor control problem, the applied voltage and residual voltage would be out of phase, causing excessive currents and motor torques that could cause power problems and damage to the motor shaft. For this and other reasons, large motors are usually started with momentary start–stop push buttons, not maintained, selector switch controls. If a power dip occurs, the motor can be restarted under controlled conditions. Maintained, selector switch motor control can be used for ride through or voltage if the shut down of the motors causes serious problems, but this should be considered only for small motors. Ride-through capability can also be provided using an undervoltage time relay in the motor circuit, this method provides automatic restart for a short period of time.

Standby and Emergency Power Supplies for Motors

There are critical motors in chemical-process facilities that on loss of power could cause a release or safety incident. Specifically, if the motor fails to operate for any sustained period of time, fires, explosions, or environmental releases are possible. Determining critical motors within process units depends on the process system. Agitators on reactors that can experience runaway reactions, situations where the process material can polymerize, fans that prevent releases, and ventilation fans that reduce the classification are examples of critical motors.

Alternate power supplies and engine-generator sets are usually used to provide emergency or standby power to process critical equipment motors. For example, plant fire water pumps have multiple backups as defined by fire codes. Diesel-driven pumps and emergency power engine sets are two of the options.

Alternate power supplies may involve the 480-volt system or a medium-voltage system, depending on where the failure is expected to occur. Selecting an alternate system requires an examination of common failure modes. If the concern is the utility supply to the plant, an alternate feed within the plant does not solve the problem. For instance, a plant experienced a major failure when a snake caused a line failure at the plant's main substation. One correction included providing a "normal" and "alternate" plant feeder system, but the same power line poles were used. If a truck hit a pole, both lines would be out. In such cases it is important to ensure that the alternate supply does not have a common failure mode with the normal supply.

Alternate systems usually use an automatic transfer switch to transfer load from the normal to the alternate supply. The controls sense the loss of power and switch from normal to alternate. The transfer time for mechanical switches is usually less than a few seconds, which is acceptable for motor application. In fact, it is desirable to allow a deliberate delay before reenergizing motors because of the residual voltage of the motors, as we previously discussed.

Standby Engine-Generator Sets

Standby engine-generator sets with automatic transfer switches and the associated controls are used to supply emergency and standby power to start and operate critical electrical motors (see Figure 6–10). The generator controls sense the loss of power and crank the engine system to cause it to accelerate to operating speed. This requires anywhere from 8 to 15 seconds, depending on the size of the unit. When the generator output voltage and frequency are at rated values, the automatic transfer switch transfers load from the normal supply to the engine-generator set, which starts and runs critical motors and other loads. Engine-generator sets are also used to provide standby power to a UPS, emergency lighting, fire-fighting equipment, and other critical loads.

The engine type can be diesel, gasoline, or liquidfied petroleum gas (LPG). Gasoline engines are usually limited to smaller size units, typically

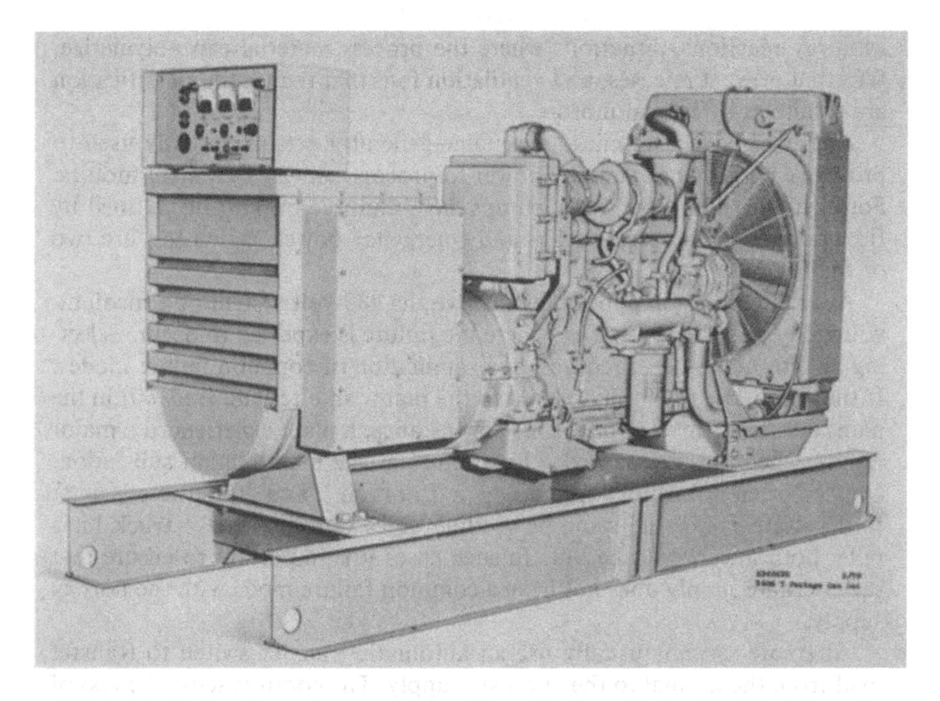

FIGURE 6-10. Standby engine-generator set (Caterpillar,Inc., Model 3406 Diesel Generator Set rated at 350 kilowatts.) The left end of the set is the generator covered by a housing that enclosed the generator and its controls. At the top of this housing is the metering and control panel for the generator and engine set. On the right side of the photo, is the diesel engine. It includes the engine, its accessories, including the starter motor and, fan-cooled radiator. Not shown in this photo are the fuel tank, automatic transfer switch, batteries and battery charger, resistance grounding panel, and the interconnected cabling. (Fabick Power Systems Co., St. Louis, Mo.)

less than 100 kilowatts. They involve a greater hazard than other types of engines because of the storage and handling of the gasoline, usually require more maintenance, and have a shorter fuel storage life. Natural gas and LPG gas-driven engines can also be used; however, the loss of the gas supply is a consideration. For this and other reasons, diesel-electrical sets are popular for this type of service. They are rugged and reliable if properly applied and maintained. They are supplied in standardized equipment and control systems, but these are options. There are a number of control systems involved in the operation of engine sets.

The governor controls the engine speed under varying load conditions. Two types of governors are available: the droop type and the isochronous type. The isochronous controls the speed within 1% under varying load conditions. The engine speed determines the frequency of the power from the ac generator. Where frequency control is important, for instance, in computer and electronic loads, the isochronous governor should be specified. The fuel tank can be provided as an integral part of the engine and can be sized for many hours of operation. Air supply, exhaust, fuel systems, and noise control are all mechanical features that require attention. On loss of power, the engine is started by a dc motor powered by a charger and battery. Monitoring the condition of the battery charger and battery is essential to ensure starting. Most control systems do this automatically. The output voltage of the generator is controlled by a voltage regulation by exitation of the dc field current of the generator. The generator field should be brushless to eliminate brush failure and maintenance.

Standby power diesel-electrical sets are typically supplied with an automatic transfer switch to transfer load from normal to generator power and back again. Control system operation is standardized, but new microprocessor-based systems are now available that provide improved diagnostics and are more reliable than earlier systems.

Control system operation typically consists of the following:

- Normal power supply voltage is monitored by undervoltage relays on each phase, which, after a time delay, signals the engine starting motor dc circuit. The undervoltage trip point and time delay are adjustable. The time delay is used so that the control system will ignore transient voltage dips. After the generator voltage has stabilized, the automatic transfer switch switches the load to the generator. After normal power returns, another timer ensures that the engine set operates a minimum time to reach operating temperature. A final timer runs the set after retransfer to exercise the unit.
- Testing the engine set periodically is essential to reliable operation. This should include occasional load testing.
- Motor starting determines the size of the generator set if a motor is the largest load. The set could be operating within its connected kilowatt load but fail to start a large motor. A rule of thumb is that the largest motor horsepower should not exceed 40% of the generator kilowatts. On a recent project, a 500-kilowatt generator was purchased to operate a 300 horsepower motor without considering the motor starting condition. Even though the operating load was within the generator rating, the motor failed to start. To correct this deficiency, an ac adjustable speed drive was added as a reduced voltage starter. In general, a motor starting study is required for each application, including voltage dip and starting torque calculations.

"Clean" Power for Process Control Electronics

Process control electronics, including process computers, distributed control systems, "smart" transmitters, PLCs, and analyzers, are vulnerable to power supply abnormalities that can cause critical systems to misoperate, shut down, lock out, and even experience component circuit board failures. Process safety is at stake when these systems malfunction. In some extreme cases, the credibility of the control system may be questionable in the minds of control room operators and manufacturing personnel.

Power outages and large voltage dips and surges would be expected to affect these systems in the same way they affect motors, lights, and other electrical loads. Process control electronics, however, are also sensitive to very small variations in the magnitude and wave shape of the voltage at its supply terminals. Even transient blips can cause microprocessor chip memories to lose information. Short-term transients of millisecond duration can couple into the process control system through the microprocessor power supplies, wiring, and system grounding. Modern electronic equipment is smaller, faster, and operates at lower power levels and is therefore more sensitive to power aberrations than its discrete component predecessors. These problems are referred to as "dirty" power problems, and an entire industry of various types of equipment has developed to solve "dirty" power problems and produce "clean" power.

The types of equipment available to solve dirty power problems, listed in terms of increasing cost and complexity, are

Isolation transformers
Line voltage regulators
Line conditioners
Engine-generator Sets
Motor-generator Sets
Uninterruptible power supplies

Each of these types of equipment has its own advantages.

The agonizing aspect of the dirty power problem is the possibility of providing equipment intended to cure dirty power problems that does not solve the problem. The solution may involve proper grounding, wiring practices, and/or process control equipment design. Some equipment may be more sensitive than others. The sensitivity of computer systems to power system aberrations has been studied by organizations, with the general conclusions that for most equipment, the acceptable voltage limit is a function of the duration of the voltage variation. Unfortunately, voltage variations on some power systems under some conditions can fall outside this envelope and cause misoperation. For a given equipment, these limits should be known. Vendor application data should always be followed and any application concerns referred to the vendor's engineering department.

Power supply abnormalities can occur for different reasons. Abnormalities are variations in the magnitude and wave shape of the voltage at the equipment terminals. Equipment failurer, faults, switching, storms, or accidents can cause distortions in the power that energizes critical process control systems. These distortions can affect microprocessor electronics. It is the extreme sensitivity of control systems and consequences of their failure that is critical. A very significant factor is an industry change in the type of DC power supply used in computer and microprocessor equipment from traditional power transformer coupled design to switching type DC supplies. This change significantly increased the sensitivity of control systems to input line power distortions. Direct current power supplies are used in all types of electronics equipment to provide operating voltages for various transistor and microprocessor circuits, therefore the problem is widespread. DC power supply quality is a significant factor in the operation of microprocessor equipment. Too often the reliability of the system is jeopardized by the quality of the DC power.

Both of these types of supplies can use power isolation transformers and output filter capacitors that provide power dip ride-through capability. Switching type power supplies operate on a different principle, utilizing switching transistors. These types of supplies are popular in computer equipment because they are smaller, more efficient, and less expensive; however, they have less immunity to ac input power variations and inject power harmonics back into the ac input power because of the power switching action of the chopper. These harmonics can react with the power system, especially capacitors, causing failures and overheating.

In selecting the proper equipment and designing a system to protect process control systems from "dirty" power problems, it is necessary to assess the vulnerability of the system to power problems and the likelihood the power system will have problems. For an existing system, it is possible to measure and record the magnitude, frequency, and duration of disturbances. This necessary instrumentation is readily available; since it is specialized and used infrequently, it may be wise to lease for a given power survey. In that case, the manufacturers of the equipment can also assist in the interpretation of results. Power records on outages, failures, and problems (or lack of problems) with existing microprocessor electronics are useful in the evaluation.

Note that even though the existing system may seem trouble free, it is possible additions may cause problems. Large adjustable speed drives that generate harmonics, large motors that cause voltage dips, UPS and switching power switches that generate harmonics into the system, and power semiconductor heater controls that produce a pulsating load on the power system are examples of problems that have been added to systems.

Separation of Systems

Many power quality problems occur because of a coupling from a common element in the system. The element could be a transformer, generator, neutral conductor, power conductor, the same conduit and cable tray, or any other shared item. Separation of systems, therefore, is a way to minimize problems. There is, of course, a practical limit: Many systems share the same power feeders, breakers, or MCC; nevertheless, steps can be taken to provide separation. One of the easiest choices is an isolation transformer.

Isolation Transformers

An isolation transformer provides a number of good features at minimal cost. This transformer has a separate primary and secondary winding (an auto transformer is not an isolation transformer); a metallic shielding can be placed between the primary and secondary winding to filtering primary power noise. In addition, the isolation transformer provides a separately derived grounding system close to the load, thereby reducing the exposure to ground noise. Also, the isolation transformer is usually a smaller transformer, which will limit any short circuit current. Transient overvoltage protection can also be provided on the primary and secondary circuits. Finally, a special type of transformer called an ultraisolation transformer provides a high degree of filtering and shielding.

Line Voltage Regulators

Acceptable limits of voltage regulation for most industrial electronic equipment is typically 10% of nominal. For a particular electronic controller, the limits depend on how much variation there is and for how long it occurs and should be verified with the equipment manufacturer. For any given power system, voltage varies as loads are added or other conditions occur.

The degree of loading on a power system exerts a strong influence on voltage regulation and other problems. The impact of load and other changes is stronger on a system operating at full capacity. For electronic equipment, acceptable limits are related to the dc power supply design, the limits of operability of the microprocessor logic and control circuits, and the other line-powered components.

If the voltage is consistently high or low, it can be corrected by changing the ratio taps on system transformers. Transformers should be supplied with ratio taps so the secondary voltage can be raised or lowered as required. This change, however, does require a shut down.

Capacitors can also be added to compensate for voltage changes. They provide an improved power factor and release system capacity. They should, however, be added with the precaution that if harmonics are present, they can produce resonance under certain conditions, that can produce overvoltage and component failures.

Line voltage regulators are also available to provide steady-state regulation within the limits of the voltage to process computer and microprocessor controls. They include variable ratio transformers, induction regulators, tap switching regulators, and ferroresonant transformers. These devices are designed for steady-state voltage control and in some cases are inherently slow and do not respond to rapid changes in voltage. Autotransformer equipment does not provide total isolation.

The ferroresonant transformer is a popular voltage regulator. It can be used individually or in conjunction with a UPS or other standby power equipment. It provides isolation and voltage regulation via a tuned winding and capacitor combination. It is a simple, reliable device without moving parts and can provide good regulation. It does, however, have limited motor-starting capability, an output voltage that can be square wave (which may produce problems with some loads), and sensitivity to line frequency deviations, which could occur if the unit were energized from an engine-generator set.

Line Conditioners

Line conditioners are a combination of voltage regulators and isolation transformers, thereby combining the features of both: line voltage regulation, isolation, and noise filtering. Transformers, voltage regulators, and line conditioners have limited, if any, energy storage capability and therefore limited, if any, capability to ride through voltage dips. Some specialized versions provide ride-through capability to facilitate the operation of transfer switches without power interruption.

Engine-Generator Sets

Engine-generator sets, motor-generator sets with flywheels, and UPSs can store energy to provide power over a sustained period of time. Engine-generator sets are limited only by their fuel supply; they can provide power for hours or days. They are usually not used as a stand-alone supply for process control systems because of the delay during the engine-cranking period, which can vary from 6 to 15 seconds, depending of the size of the system. Most process control systems could not tolerate this interruption. Engine sets are typically used as a long-term source of power for UPS systems and to provide power to lighting motors and other critical loads. UPS systems derive their power from batteries when they lose their input ac power. The battery capacity is typically 15 minutes to allow for an orderly shut down if power to the unit is not restored.

The energy storage capability of a motor-generator set depends on the inertia of the shaft and is usually less than ½ minute.

Motor-Generator Sets

Motor-generator sets consist of an ac motor coupled to and driving an ac generator. The stored energy is a direct function of the inertia added to the shaft. The speed of the standard induction motor drops off 1–3% with load; since the frequency of the power developed by the ac generator depends on speed, the output frequency will drop off proportionately. To provide precise output frequency, a synchronous motor can be used instead of an induction motor. A specialized version of the MG set using a dc motor instead of an ac motor and batteries instead of shaft inertia can provide additional ride-through power, depending on the capacity of the batteries. Motor-generator sets provide isolation from power line dips or surges and are used as stand-alone devices or in conjuction with UPS systems to provide ride through during switching. Motor-generator sets also provide pure sine wave power, which UPS systems do not.

Uninterruptible Power Supplies

Uninterruptible power supply systems, which provide no-break power, are used to provide standby power to process control. Using solid-state switches, load transfers occur in less than a quarter of a cycle; in some situations, there is no break at all. UPS systems synthesize ac power from dc power. The system consists of a rectifier-battery charger, battery assembly, static inverter unit, and static switch (see Figure 6–11). The rectifier-battery charger converts ac input to dc power to charge the battery assembly and provide load dc power to the inverter; the inverter converts dc power to ac power by chopping up the dc using power semiconductors, transistors, or silicon controlled rectifiers (SCRs). The output wave form is an approximation to a sine wave. It can be a stepped sine wave or a series of pulses. The types of inverters that produce these wave shapes are called six-step or pulse-width modulation (PWM) inverters. These wave shapes have high-frequency harmonic power components that can interact with microproces-

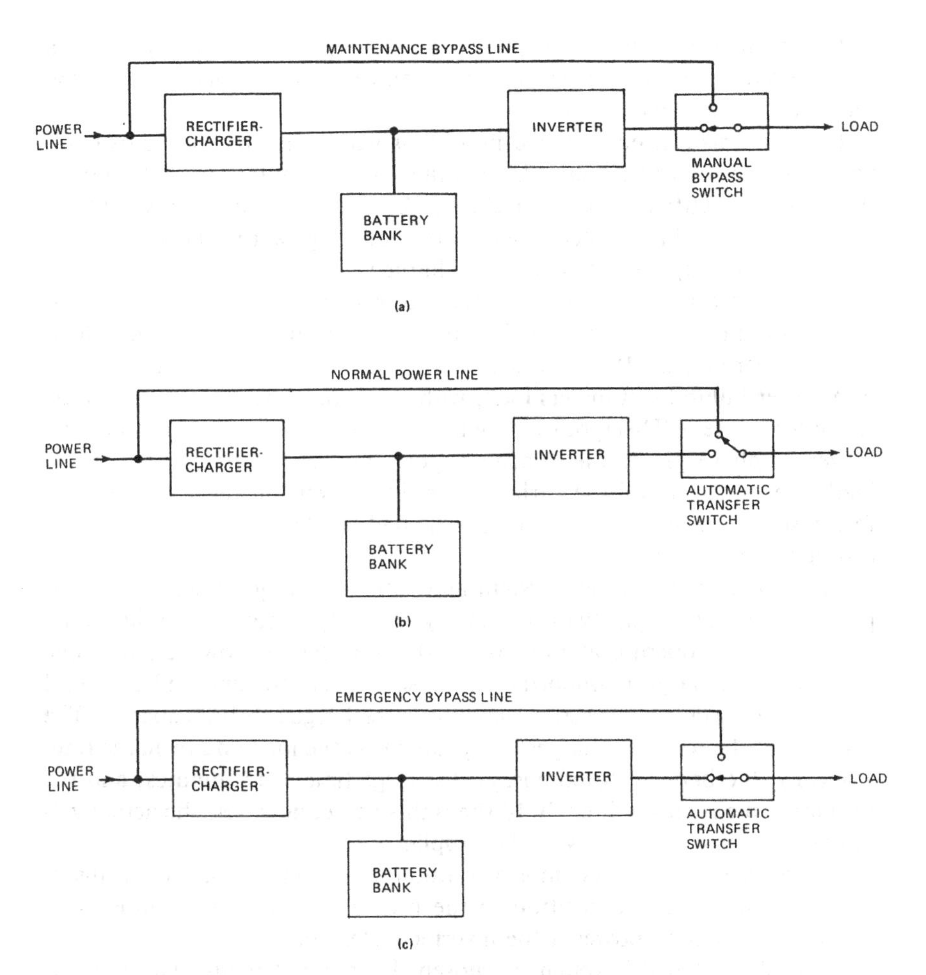

FIGURE 6-11. Three configurations of a UPS system: (a) manual system; (b) forward transfer mode; (c) reverse transfer mode. Each system has advantages and disadvantages that need to be reviewed with the manufacturer. (Dranetz Technologies, *How To Correct Power Line Disturbances* TP-103957-TA.)

sor loads, especially those that have switching dc power supplies. UPSs are usually specified as having a maximum of 5% total harmonic distortion. If, however, the load is largely switching power supplies of the UPS is fed from a generation, harmonic problems can occur. The type of loads and supply should be described to the UPS supplier. The static switch operates only when a failure occurs, but it does so within a quarter cycle, which should not disrupt microprocessor electronics. One of the early applications of UPS systems was to provide power to critical process burner systems. In these systems, the static transfer switch had to be fast enough not to drop the sensitive interlock relays.

UPS systems can be arranged in various ways, each having its own advantages and disadvantages. The continuous UPS system provides power through the rectifier charger to charge the batteries and supply load. If input ac power is interrupted, the battery supplies power to the load. When power is restored, the batteries are recharged.

The forward transfer system is not a continuous system. It is provided with a static transfer switch, which, in its normal position, connects load to the power system. If the static switch control system senses low voltage or a power failure, it transfers load, with a bumpless transfer, to the inverter battery system. This type of system is sometimes referred to as a standby power supply (SPS). In the "normal" position, the process control system loads are subject to all the variations in line power; however, MG sets or ferroresonant transformers can be provided in this line to provide ride through and regulation.

The sizing of the battery system and rectifier-charger system is an important issue. The capacity of the battery assembly is indicative of how long the system can operate at full load without recharge; however, the time required to recharge is important because if it is too long and a second interruption occurs, the system may not be recharged to full capacity. The capacity of the rectifier-charger is dependent on the load and recharge rate. Typically, the battery stored energy discharge time is 15 minutes, and the recharge time is approximately 10 times this value; however, the actual values need to be confirmed with the supplier.

The reverse transfer system is a continuous system with a static transfer switch, which is in the position to the inverter in the normal mode and transfers to standby power if the inverter system fails.

Regardless of which system is chosen, bypass and maintenance circuit breakers or switches should be considered for manual operation and to isolate the equipment for maintenance.

Harmonics

The impact of power line harmonics on all parts of the system must be considered. Harmonics can cause systems to fail, overheat, and misoperate. Harmonics are caused by switching power semiconductors, which circulate back through the power system. Switching dc power supplies in microprocessor equipment can be reflected back into the UPS system or normal power when the UPS is in by pass. It is essential to communicate the harmonic load requirements to the UPS manufacturer. Failure to do so has caused serious problems with this type of equipment. UPS systems generate harmonics into their supply; if this includes an engine-generator set, the vendor of equipment must be informed to ensure that the generator controls are compatible with harmonic sources.

CABLE SYSTEMS FOR CHEMICAL-PROCESS FACILITIES

Cabling systems in chemical-process facilities interconnect the electrical power equipment and the electronic process control systems. They provide power and control to motors, heaters, lighting, instrumentation, PLCs, microprocessors, and process computers within the system. The operability, reliability, and safety of the process depends on these systems. The safety issues relate to the reliability of the systems, protection of people from high voltages by cable shield or metal armor, grounding, and protection of people and property from toxic gases and fumes that can occur from cable fires.

Complying with the *NEC* requirements concerning the types of cables and their application is the proven way to meet safety requirements. UL, NEMA, and IEEE standards provide additional information; use UL-listed and marked cables if possible. Cable manufacturers and trade associations are also excellent resources. Cable manufacturers can provide valuable application information; in particular, they can provide data concerning chemical compatibility of cable insulations. Chemical compatibility should always be verified before a particular cable insulation is selected.

Cabling systems can be categorized according to the voltage on the conductors. The systems include

Medium-voltage power (more than 1000 volts)
Low-voltage power (to 1000 volts)
Instrumentation (low energy milliampere, millivolt)
Computer and microprocessor (digital signal voltages)

Power cables typically consist of three power conductors, one per phase, with or without equipment grounding conductors, which can be bare or insulated conductors. The individual conductors are insulated, and the entire assembly is covered by an overall protective jacket. If equipment grounding conductors are provided, they have a smaller cross-sectional area than the phase conductors. The minimum acceptable size of grounding conductors is indicated in the *NEC* table 250-95 and the UL 1277. Grounding conductors with the cable provide a safer equipment grounding system, especially for larger size cables. It is still, however, required and essential that cable tray, conduit, sealtite, and so on, are grounded to provide an effective (and safe) equipment grounding system.

The overall protective jacket can be metallic or nonmetallic. At more than 2000 volts, a metallic shield is provided to confine the high-voltage electric field to within the cable. This is essential to protect people and the cable from high-voltage ionization or corona that could occur outside a high-voltage cable. Terminating the shield at each end of the cable requires specialized cable stress cone termination devices and good workmanship. If it is done incorrectly, even the slightest imperfection can cause a failure to occur, perhaps a few months after the installation is complete. A major plant shut down was caused by a failure of a high-voltage connection due to a small air bubble in the dielectric material. After the work is complete, high-voltage installations should be tested by following IEEE standard test procedures.

Single-conductor power cables can be used instead of three-conductor cables, but there are disadvantages. Single conductors do not confine the electromagnetic field as well, and they are subject to greater forces during a short circuit. An incident that occurred at a plant illustrates this point. A short circuit occurred in switch gear as it was experiencing maintenance. The single-conductor cable feeding the faulted phase broke loose from the cable tray due to short-circuit forces, flew into the air, and landed on a car in the parking lot. This would not occur if three-conductor cable were used.

Low-voltage power and control cable can be more than three conductors depending on the control requirements. Nineteen conductors is a practical limit for the number of conductors in a control cable. Individual conductors are usually stranded and identified by the color of the insulation or by numbers according to NEMA standards. The degree of stranding determines the flexibility of the cable; in situations where flexibility is required, special stranding can be specified.

The conductor material can be copper or aluminum. Copper is usually favored in smaller sizes; aluminum is competitive in larger sizes. That is, aluminum has a cost and weight advantage over copper in larger sizes, but aluminum requires a larger cross-sectional area than copper for the same current rating. Aluminum conductors require special precautions and care in terminating and connecting the conductor because of the thermal expansion characteristic and the formation of oxides. The reliability of aluminum terminations is good where established industry practices concerning the preparation of the conductor and the connection have been used.

Compression-type connectors should be used. A cabling system consists of the cables and the raceway, cable tray, or other method to support, enclose, and protect the cables. Raceways include conduit, flexible conduit, and other systems as defined in the *NEC*. Cable systems can be run overhead or in underground duct systems or be direct buried. The *NEC* specifies minimum burial depths. Underground systems are typically used where above-ground systems are at risk to mechanical damage, but they are subject to underground damage from spills, in particular, underground excavation work. A number of plant shutdowns have occurred when underground excavation resulted in damage to underground electrical circuits. Underground electrical conduit or buried cable systems should be marked with warning signs. It is common practice to color all underground concrete conduit red or to put red dye in the concrete. Underground systems are sometimes used at main plant substations as well as for distribution within the plant.

Cables are sized according to their cross-sectional area based on the current in the power conductors, the temperature rating of the conductor insulation, the heat-conducting media around the conductors, voltage drop, and other factors.

Conductors are usually stranded for flexibility. The *NEC* requires stranded conductors in sizes larger than No. 8. The *NEC* also indicates the minimum bending radii for conductors, and this dictates the space required to terminate large power conductors in boxes.

The conductor current-carrying capacity, or ampacity, for the various sizes, types, and conditions are found in Article 310 and Appendix 310B of the *NEC*. The conditions of use determine a conductor's ampacity. Some of the conditions that determine how much current a given conductor can carry are whether the conductor is in free air, buried underground, or in a conduit with other cables.

Several types of code-recognized, UL-listed cables are used in chemical process facilities, including

Type TC (tray cable)—Low-voltage power and control
Type PLTC (power-limited tray cable)—Instrumentation signals
Type MV (medium-voltage) High-voltage cable
Type MC (metal-clad) Cables

Cable tray systems are popular in chemical-process industries. Representatives of the Chemical Manufacturers Association who were members of *NEC* panels were responsible for the acceptance of cable trays, and their work is reflected in Article 318 of the *NEC*. Cable trays can be used in almost any type of process, but they cannot be used where they would be exposed to mechanical damage or in hoistways. They have been used extensively in chemical-process facilities without major problems regarding mechanical damage. They provide lower cost cable installations because of the relative ease with which the tray systems can be installed and cables added compared to conduit systems. They are also flexible. Cables can be added or removed relatively easily. When cables are pulled through conduits, they can be damaged, especially if the conduit systems have a number of bends.

The maximum number of cables that can be installed in a conduit is limited by the *NEC*. Article 318 also limits cable tray fill. Tables 318-8 and 318-9 indicate the fill-in square inches for various tray widths. In general, cable tray fill should not exceed 40% of the cross-sectional area of the trays. Overfilling cable trays is especially a problem at control and electrical rooms, where the cable density is the highest. Allowing extra cable tray capacity in these areas to facilitate additions and ease of installation is prudent. Both cable trays and conduit are good equipment ground conductors. They must, however, be bonded together effectively back to the power transformer.

Both cable trays and conduit can be constructed of different materials to meet the severe environmental conditions in chemical-process facilities. Cable trays can be constructed of aluminum, galvanized steel, plastic-coated aluminum, steel, or fiberglass-reinforced polyester resin. Metallic cable trays can be used as equipment-grounding conductors. They are effective, especially aluminum, because of the large cross-sectional area. If they are used as an equipment-grounding conductor, they must be labeled as such, and the cross-sectional area of the cable tray metal sections must be marked on the cable tray.

A safety concern with the use of all conduit systems is the possibility of the conduit acting as a pipe to channel process fluids into a control or electrical room. This has occurred when conduit was connected to a thermowell, canned pump, or other process connection, and the process seal failed, pressurizing the conduit system. An incident like this occurred at an LNG facility when the process pump seal at the motor leads failed and pressurized LNG was forced through the motor terminal box, through an explosion seal in the conduit into the box, and through 200 feet of underground conduit into an electrical substation building.

The construction of the cable needs to be specified to provide a rigid support system for the particular cable loading and span between tray supports. NEMA standard VE-1-1984, *Metallic Cable Tray Systems,* provides guidance in specifying the correct class of tray for the particular application. This design should use appropriate safety factors and only cable tray types with a proven record of durabilty.

The contribution of cable insulations to the propagation of flame and the products of combustion in a fire is a safety issue that has received a great deal of attention recently. New York state recently passed a law requiring the testing of building materials for toxicity. Nonmetallic electrical conduit and electrical wire insulation were tested. The concern is the contribution of electrical cable and other items to the products of combustion.

Tray cable has been tested using IEEE test standard 383 to determine its acceptability concerning the propagation of flame. The test, called the vertical flame test, uses burners to ignite the cable in a vertical section of trays and evaluates the flammability of the cable. Cables that are UL listed as tray cables must pass this test.

NEC articles 300–21, "Spread of Fire or Products of Combustion," and 300–22, "Wiring in Ducts, Plenums, and Air-Handling Spaces," are also concerned with this issue. Section 300–21 addresses the issue of electrical penetrations through fire-rated walls and requires firestops for conduit, cables, or cable trays that pierce these fire-rated walls. Several devices and methods are available. Some involve pouring or foaming a grouting-like material to seal the opening; others use mechanical grip-type devices. They should be UL tested and listed as capable of maintaining the fire rating of a wall.

Article 300–22 addresses wiring in ducts, plenums, or other spaces used for environmental air and is concerned with the wiring systems that can contribute smoke and other products of combustion to air systems people may breathe and contact. The wiring systems must be enclosed in metal conduit or metallic tubing; if cable is used, the insulation material must be such that the cable meets industry requirements for safe emissions during a fire.

Designing cable systems on a system basis with separation and isolation of systems can minimize the potential for cable fires and the extent of damage if they do occur. Wherever possible, separate high-voltage from low-voltage systems, instrumentation from other systems, and computer cabling from other cables. This practice will also simplify installation of the cables and will reduce electrical noise problems in instrumentation and computer cables.

CONCLUSION

This chapter discussed and developed the many aspects of electrical safety in chemical processes, in particular, the following:

Electrocution and personnel safety
Lightning protection
Static electricity safety
Circuit and equipment protection
Enclosure selection

Electrical power system reliability and quality
Cabling systems

The principles developed are discussed at great length in the *NEC,* IEEE, API, and NEMA standards and articles and papers referenced at the end of this chapter.

The discussion of electrocution indicated that very small electrical currents can have a devastating effect on the human body. Personnel protection is provided by guarding energized electrical parts, selecting electrical enclosures that are protected against water and corrosion, providing adequate insulation and working space, grounding, using ground-fault circuit interrupters, and providing system designs that eliminate or reduce the need for hot work.

Lightning protection of buildings and structures is achieved by the system of aerial terminals, down conductors, and grounding electrodes, which attract and divert the strike to earth. Storage tanks and other vessels on the top of structures that contain flammables must have a thickness of at least 3/16 inch and be properly connected to earth.

Static electricity is generated in many process situations, but the accumulation of static charge and generation of an ignition-capable spark usually requires an isolated ungrounded metal object. Static charge accumulation can be quite significant if plastic piping and vessels are used.

Electrical enclosures located in chemical-process areas and not watertight or corrosion resistant are a major factor in electrical system failures. NEMA 3 or 4 enclosures are usually good choices in unclassified locations; however, the corrosive properties of the particular chemicals involved should be considered when selecting the enclosure materials of construction, particularly, the enclosure gasketing.

Selecting the proper type and size of fuses and circuit breakers is essential to ensure a safe installation. Fuses and circuit breakers must be used within their interrupting rating and must limit the energy in any fault to within the withstand rating of the downstream equipment. Analysis of electrical systems under fault conditions can be done using available computer software.

Grounding is involved in lightning protection, static electricity safety, electrical equipment and wiring system protection, and protection of computer and instrument systems against electrical noise. The Ufer ground, consisting of the interconnected building foundation rebar, is effective as a grounding electrode for lightning protection and computer grounding, especially in poor soil conditions.

Resistance grounding of the transformer neutral for electrical wiring system grounding has the advantage of providing a stable wiring system ground without causing a fuse or breaker to operate on the first ground fault.

Equipment grounding is the interconnection of all metallic enclosures, cable trays, conducts, or boxes back to the supply transformer. The system neutral, or grounded conductor, should be connected to earth or building steel at only one point. Neutral grounding at more than one point can cause stray currents to flow in building steel or piping, which can interfere with microprocessor electronics. Ground inspections are necessary to verify that the system is interconnected.

Power system reliability and quality are important safety and environmental issues in chemical processes. The loss of lighting, critical process motors, and controls and instrumentation during a power upset can have undesirable effects on chemical processes. Emergency lighting should be provided at critical locations. Standby power for critical motors can be provided by a Diesel Electric Generator set or by switching to an alternate feeder and supply.

The inplant power distribution system can be designed as either primary or secondary selective. The amount of redundancy should be commensurate with the risk of an outage and the process consequences. Alternate feeders and switches should also be provided to minimize the necessity of hot work on energized electrical equipment during maintenance.

Instrument and control power systems should be protected so power line abnormalities do not cause process upsets or shut downs. There are a number of choices as to the type of equipment and system design for each application. Identifying what protection is required and whether the selected system accomplishes the goal are important. Complicated back-up systems can cause outages due to their own failures. The power distribution systems for instruments and controls should be designed in a distributed fashion so no one failure can cause a complete shut down of a reactor or process system.

Cable systems should be engineered for ease of installation, flexibility, and working space for expansion. Cable types and design should comply with *NEC,* UL, NEMA, and IEEE standards. Cable trays provide a system that gives significant advantages over conduit systems, but system design must comply with Article 318 of the *NEC.*

REFERENCES

American Petroleum Institute. March 1982. *API RP2003. Recommended Practice for Protection Against Ignitions Arising Out of Static, Lightning, and Stray Currents,* 4th ed.

American National Standards Institute/Institute of Electrical and Electronics Engineers. 1986. *ANSI/IEEE Standard 141. Recommended Practice for Electrical Power Distribution for Industrial Plants.*

Biddle Instruments. April 1981. *Getting Down to Earth: Manual on Earth-Resistance Testing For Practical Man.* Blue Bell, Penn., 4th ed.

Britton, L.G., and J.A. Smith. 1988. Static Hazards of Drum Filling. *Plant/Operations Progress* 7 (1, Jan.):53–63.

Bussmann, Cooper Industries. 1990a. *SPD Electrical Protection Handbook.* St. Louis, Mo.

Bussman, Cooper Industries. 1990b. *NE90 Overcurrent Protection and The 1990 National Electrical Code.* St. Louis, Mo.

Carpenter, R.B. Jr. Lightning Elimination. IEEE Paper PCI-76-16-Conf Record 76CH1109-8-IA.-1976

Daly, J.M. Gas/Vapor-Tight Cables For Class I Hazardous Locations. IEEE Paper PCI-81-10.-1981

Daly, J.M. 1983. NEC Cable Types and Applications. *IEEE Transactions on Industry Applications* 1A-19 (6, Nov./Dec.).

Daly, J.M. 1990. Flammability and Toxicity Testing of Wire and Cable. *IEEE Transactions on Industry Applications* 26 (3, May/June).

Dalziel, C.F. 1972. Electric shock hazard. *IEEE Spectrum* (Feb.):41–50.

Designing Reliable Power Systems for Processing Plants. 1989. *Electrical Systems Design* (Jul.):36–40.

Eichel, F.G. 1967. Electrostatics. *Chemical Engineering* (Mar.):153–167.

Expert Commission for Safety in the Swiss Chemical Industry (ESCIS). 1988. *Static Electricity: Rules for Plant Safety* 7 (1). *Plant/Operations,* Vol 7, No 1 Jan 1988.

Freund, A. 1988. Nonlinear Loads Mean Trouble. *Electrical Construction* (Mar.):83–90.

Hoffman Engineering Company. 1989 *Electrical Electronic Enclosures & Equipment: Specifiers Guide.* Anoka, Minn.

IEEE. Cable Systems. *IEEE Standard 446: IEEE Recommended Practice for Emergency and Standby Power Systems for Industrial and Commercial Applications,* Chap. 8.

Journal of The Franklin Institute—Special Issue on Lightning Research 283 (6, June 1967) Philadelphia, Pa.: Franklin Institute of the State of Pennsylvania.

Kelly, L. J., A. J. De Chiara, and J. R. Cancelosi. Tray Cable Selection in Petroleum and Chemical Plants. IEEE Paper PCI-81-1.

Klinkenberg, A., and J.L. Van Der Minne. *Electrostatics In the Petroleum Industry.* A Royal Dutch/Shell Research and Development Report. New York: Elsevier Publishing Co. 1958

Kouwenhoven, W.B. 1969. Human Safety. *Electrical Safety Practices.* ISA Monograph 112.

Lee, R. H. 1979. Lightning Protection of Buildings. *IEEE Transactions on Industry Applications* Ia-15 (3, May/June):236–240.

Lee, R.H. 1982. The Other Electrical Hazard: Electric Arc Blast Burns. *IEEE Transactions on Industry Applications* IA-18 (3, May/June):246–250.

Mages, L.J. 1987. Test-Running Standby Engine-Generator Systems. *Plant Engineering* (Jan. 22).

Nash, H.O. Jr., and F.M. Wells. 1985. Power Systems Disturbances and Considerations of Power Conditions. *IEEE Transactions On Industry Applications* IA-21 (6, Nov./Dec.).

National Fire Protection Association. 1983. *NFPA 77. Recommended Practice on Static Electricity, 1983*. Batterymarch, Mass.

National Fire Protection Association. 1988a. *NFPA 110 Standard for Emergency and Standby Power Systems.*

National Fire Protection Association. 1988b. *NFPA 101 Life Safety Code.*

National Fire Protection Association. Article 700, Emergency Systems. *NFPA 70 The National Electrical Code.*

National Fire Protection Association. Article 701, Legally Required Standby Systems. *NFPA 70 The National Electrical Code.*

National Fire Protection Association. Article 702, Optional Standby Systems. *NFPA 70 The National Electrical Code.*

Owens, J.E. Static Electricity. ISA-1964 *Electrical Safety Practices* Monograph 110:113–127.

Palko, Ed, ed. 1987. Selecting and Applying Standby Generators. *Plant Engineering* (Apr. 23).

Parker, J.B. Cable and Conduit Sealing Requirements in Fire Rated Barriers. IEEE Paper PCIC-83-24. 1983

Shield, B.H. 1967. Battery Failure Causes Major Failure of Gas Turbine. *Loss Prevention, A CEP Technical Manual,* Vol. 1. AICHE.

Towne, H.M. 1956. *Lightning: Its Behavior and What To Do About It.* Thomas Lightning Protection, Inc., St. Paul, Minn.

Underwriters' Laboratories. 1969. Electric shock As It Pertains to the Electric Fence. *Underwriters' Laboratories: Bulletin of Research.* (14, Dec.)

UPS. Solving the Computer Power Problem. 1984. *Emergency Power Engineering Inc.* (Nov.) Costa Mesa, Calif.

U.S. Department of Labor. 1983. *Controlling Electrical Hazards.* OSHA 3075.

U.S. Department of Labor. 1985. *Ground-Fault Protection on Construction Sites.* Rev ed. OSHA 3007.

7

Measurement and Final Control Elements

The safe operation of chemical-process facilities is strongly dependent on process data provided by measurement instrumentation and control of the process by the operation of final control elements.

Measurement instrumentation provides input data on the state and condition of the process, typically of temperatures, pressures, levels, flows, and so on, to process controllers, alarm, or display systems located in the control room. Typical measurement systems include thermocouples, resistance temperature detectors, differential pressure (d/p) cells, orifice plates, magnetic flow meters, vortex meters, weigh cell systems, and any associated transmitters. Transmitters convert low-level signals, like the millivolt output from a thermocouple, to a standard 4–20 milliampere dc signal to the control room instrumentation. Final control elements consist of valves, pumps, agitators, fans, process heaters, and other equipment that directly changes the process. Whether the control room control system is a modern distributed control system, individual analog or digital controllers, PLC, relay, electromechanical drum programmers, or other control and display systems, the system is only as safe, reliable, and nonpolluting as the measurement and final control elements. The most sophisticated control scheme in the world cannot compensate for poor measurement and final control elements. If the temperature measurement is inaccurate or unreliable, overheating is possible. If a weigh system signal understates the batch weight charged to a reactor, a runaway reaction can occur. If a control valve fails to close, a process upset is possible.

INSTRUMENTATION FLOW DIAGRAMS

The first and most important step in the design of measurement and final control elements is the development of the instrumentation flow diagrams. Some companies refer to these as piping and instrumentation flow diagrams or engineering flow diagrams. These diagrams show the vessels, reactors, tanks, and piping in diagramatic form with the measurements and final control elements. The symbology varies within various companies, as does the degree of detail. The diagrams indicate what is to be measured (temperature, level, pressure, flow, etc.), where it is to be measured (middle of distillation tower, side of accumulation), failure mode of valves, instrument identification numbers, and general loop designation (level in tower controls bottoms flow). They do not, in general, distinguish the types of measurement elements except that they indicate orifice plates. These diagrams are key in defining the measurement and control systems for a facility and deserves a total and detailed review with process, instrument design, and other interested parties. It is essential that the process be defined in terms of flow rates, temperatures, pressures, volumes, process dynamics, limits of process variables, chemistry (including unusual conditions), characteristics of the various streams, corrosiveness, flammability, toxicity, and so on, to complete a review of these critical documents.

PROCESS MEASURING ELEMENTS

The design of process measuring consists of selecting a type of measuring element for the particular service and conditions, the type of installation, and in some cases (e.g., a pressure switch) the trip setting. Within the industry, companies, and plants, there is a vast area of experience and proven practice in the selection of the type of instrument for a given type of service. Instrument manufacturers and various ISA publications provide technical data and guidance in applying instrumentation. In the final analysis, however, it is prudent to use devices that have a proven record of reliability and durability for a given service.

The plant maintenance department has important input to the type and manufacturer of the instrument. It is always desirable to select the type of instruments the plant is trained to service, calibrate, and has spare parts for. It is also highly desirable to use the same type of instrument wherever possible and avoid a multiplicity of devices. As an example, a new batch process was added to a plant that previously had no experience with batch processes. A number of critical flow streams required batch metering. The instrument engineer selected a different type of flow meter for each stream, including a turbine meter, rotameter, magnetic flow meter, orifice plate, and oval gear meter. The types of meters could have been limited to two to minimize maintenance requirements.

The first, most important, and sometimes most difficult step in specifying a measuring or control element is acquiring the correct process data. Hix (1972) indicates that data provided by the process engineer for instrumentation are sometimes unreliable and inadequate and, further, that unsafe operations can occur if the instrumentation cannot cope with reality. Therefore, it is essential for the instrument engineer to work closely with the process engineer in defining process conditions for the measurement and control elements. This is especially important during project execution as changes occur but is also important for existing operations.

Complete instrument specifications are valuable in this respect, in that they require essential process data. The complete specification is a record for process design to ensure the accuracy of the process conditions. It is also important for the process and instrument engineer to ensure that the limitations for the particular device selected are within process operating conditions. Temperatures, pressures, flow rates, or other conditions like coating of probes, pulsating flows, and particles or foaming in the process need to be identified. Response time may also be a problem for some sensors. Special corrosive conditions should also be identified.

A particular type of measurement element can be selected based on the type of service, process conditions, experience, and established practice. The following is a brief discussion of the various types of measurement elements. A comprehensive and detailed analysis can be found in instrumentation text books, ISA manuals, manufacturers' data, instrumentation trade magazines, and from experienced instrument engineers.

Temperature Measurements

Thermocouples and RTDs have largely replaced filled systems as temperature measuring elements. RTDs are generally more accurate and stable than thermocouples. Both require thermowells or resistance bulbs to protect the element from the process and facilitate replacement or calibration with contact with the process.

Thermowells or resistance bulbs slow the response of the temperature element, which should be considered where fast response is required. Thermowells can be designed to reduce the thermal response. Transmitters are usually required to convert the millivolt thermocouple signal to a standard 4–20 milliampere dc signal; however, there are PLCs and microprocessor control room instrumentation that can accept thermocouple or RTD inputs directly without transmitters. Where transmitters are required, they can be located in junction boxes or cabinets in places where they are exposed to temperature extremes or environmental conditions. Transmitters should have input/output isolation to avoid ground loops, have adjustable zero and span, and be circuit protected against electromagnetic interference (EMI). Thermocouples should be tip grounded. They are usually provided with upscale burnout protection so if the thermocouple, RTD element, wiring opens, a high alarm or failure will be indicated. The location, orientation, and insertion length should be correctly defined to provide a valid temperature measurement. Wells should be located where the flow velocity is high, for instance, in a piping elbow wherever possible. Vertical mounting should be avoided, as should bottom entry into equipment. The thermowell material should be compatible with the process fluid, Type 304 stainless steel as a minimum. Failure fatigue can occur in gas service at high flow velocity if the thermowell is too long. The thermowell manufacturer should be consulted in high gas flow applications.

Pressure Measurements

Pressure transmitters are typically strain gauge or capacitance type with adequate overrange protection. Parts in direct contact with the process fluid must be compatible with the fluid; otherwise, a filled protective seal can be used.

Self-actuated pressure regulators can be used where exact pressure control is not required. Diaphragm rupture is a common cause of pressure regulator failure, and a downstream relief valve is recommended if the full supply pressure can damage downstream equipment. Bypass valves around regulators should be avoided. Pressure switches with adjustable setpoints should be provided with hermetically sealed contacts. Hermetically sealed contacts are acceptable in Division 2 locations without explosionproof enclosures. They also provide environmental protection for the contacts. If the process diaphragm fails, it is possible for process fluids to pressurize the switch enclosure and force process fluids through the conduit or cabling system back to an electrical or control room. In classified locations, the *NEC* requires a leakage point and a second process seal to prevent propagation. A specially designed pressure switch isolates the sensor head and switch compartment and provides a leakage point as part of the pressure switch assembly. The range of adjustability should match process requirements, and the switch should be wired to fail safe, as determined by process requirements, in the event the contact or wiring opens.

Level Measurements

Level measurement is available in a number of options, including d/p cells, which measure the pressure head of the liquid, displacement type of controllers and transmitter, ball float switches, capacitance, ultrasonic, rotating paddle, and conductivity and other types of instruments. Devices that contact the process fluid will have a serious problem with some corrosive and toxic chemical fluids. The mechanical failure of linkages in process streams should also be considered. Noncontact-type level devices, for instance, nuclear level devices, should be considered for toxic and corrosive streams. Nuclear instruments require special regulatory requirements and special safety maintenance considerations.

Differential pressure cells are a common type of level measure. Accurate level measurements depend on a consistent process material density. Another condition related to accurate measurement is vessel pressurization. If the vessel is pressurized, an equalizing low-pressure tubing connection to the top of the vessel above the highest level is required. The minimum level that can be measured is at the bottom connection of the vessel. Connections at the bottom of the vessel can accumulate solids. For some process fluids, purging the tubing sensing lines may be necessary. The bubbler or dip pipe method of level measurement uses an inert gas flow into the vessel and a d/p cell. Dip pipe pluggage can be a problem. Level measurement is usually not an accurate system and is not required to be. It is essential not to overfill a tank or vessel, and sometimes a backup level switch is used.

Flow Measurements

Flow measurement covers a significant part of the measurement systems in any chemical-process facility. Flow measurement involves consideration of more process conditions and the choice of more measurement devices than any other process variable. To select and apply a flow meter, it is essential to know the following:

State of the fluid—liquid, gas, or two phase
Flow rate—minimum, maximum, operating range
Temperature range
Fluid characteristics Viscosity
  Specific gravity
  Conductivity if magmeters are to be considered
Service—batch or continuous
Flow—Continuous or pulsating
Stream—contain particles or gas bubbles

The flow meter must match the process piping specifications and be installed in the piping system, where it will be filled with the process fluid. Special precautions are required for certain types of meters (e.g., the fluid must not attack or dissolve the liner in a magnetic flow meter). Unless so indicated, the flow meters measure volumetric flow, not mass flow.

For many years, in some companies the orifice plate–d/p cell combination was considered the norm for flow metering. The other choices included magnetic flow meters and turbine meters. Now, a wide spectrum of relatively new flow meters with significant advantages over orifice flow meters in terms of rangeability and accuracy are available. They include

Vortex meters
Coriolis mass flow meters
Improved magnetic flow meters
Ultrasonic meters
Swirl meters

The main disadvantage of orifice plate flow metering is the limited rangeability. It is usually limited to a 3-to-1 flow range, which, because of the square root flow characteristic, would result in a 9-to-1 signal span. The other main disadvantage is the tubing installation from the orifice plate to the d/p cell. This tubing can be damaged and fail, can plug, and may be required to be purged or heat traced.

The following is a brief discussion of the features of various types of flow meters. For a detailed discussion of how they work and how they should be applied, refer to *Industrial Flow Measurement* by David W. Spitzer.

Vortex meters have a wide rangeability, typically 10 to 1, can be used with gas, steam, or liquid service, and are relatively easy to apply. But like many flow meter types, they must be installed with a minimum length of straight sections of pipe upstream and downstream of the meter.

Magnetic flow meters are established meters for liquid streams that are slurries, corrosive, or viscous. They have wide rangeability and good accuracy. They require the pipeline to be full and the process stream to have a minimum conductivity; however, newer versions of the magmeter have reduced the conductivity requirements so a wider range of fluid streams can be used. Magmeters must be properly grounded, and power wiring to the magnetic field coils is required. Coriolis-type meters provide high-accuracy mass flow measurements with a minimum of maintenance. They require special precautions with respect to mounting and are excellent competition for load cells.

Turbine meters are very accurate over a wide range of flow. They sense flow by the action of a propeller-like sensor, which is on bearings. Bearing wearout is a problem, which is why they are suited to clean lubricating streams. They should be installed in the piping system with upstream strainers that prevent pipe scale or dirt from damaging the meter. The meter can also be damaged by gases in the pipeline ahead of the liquid.

Rotameters can be used on clean flow streams, but in chemical-process applications, they must be the armored tube design, not glass tube. Rotameters must be mounted vertically in the piping system with the flow up and are not well adapted to batch flow metering. Ultrasonic flow meters and doppler flow meters have the advantage of sensors located outside the piping. They are specialized meters and can be sensitive to gas bubbles or slurries. Therefore, communicating the stream characteristics to the manufacturers of these specialized types of meters is essential.

For all types of inline instrumentation, coordinating connections, sizes, and material specifications with piping specifications is essential. The minimum upstream and downstream straight piping distances are required for many types of meters. Some flow meters are sensitive to piping vibration and external forces from the piping system. Many newer flow meter designs include microprocessor electronics, so-called smart transmitters that have the capability to communicate with microprocessor systems that greatly enhance the diagnostic, ranging, and calibration capability of the meter. The microelectronics, however, can be sensitive to temperature extremes, vibration, and EMI. If the microelectronics is located in the meter body, process fluid temperature extremes and piping system vibration may cause the electronics to fail. If transmitter enclosures are located in classified locations, they may have to be explosionproof, intrinsically safe, or nonincendive. The enclosures should also be watertight, NEMA 3 or 4, and corrosion resistant to the particular environment.

FINAL CONTROL ELEMENTS

Final control elements change the process in response to control action by a process controller, interlock, or operator manual control. Final control elements can be valves, on–off or modulating control valves; the motors for pumps, fans, agitators, or compressors, on–off or speed control; or process heaters, on–off or modulating. Most of the final control elements in a chemical process unit are valves. On–off valves are used extensively in batch process to fill and dump charges and batches. They are also used in continuous processes to shut off flows and take other action. Sizing is not a concern for on–off valves except that they have minimum pressure drop when opened. Tight shut off is a concern with on–off valves. The valve inner works design is dictated by the process fluid and tight shut off. Plug and ball valves are typically used for this service, with pneumatic operators spring loaded to fail open or closed as indicated on the instrumentation flow sheet. On–off valves are usually provided with limit switches to signal when the valve is open or closed or, in some cases, both. Valve limit switches have been a serious reliability problem for batch control systems. The severe environment at the valve plus difficulty achieving a reliable mechanical coupling between the valve stem or actuator and the limit switch arm or actuator is the reason for poor limit switch reliability. The limit switch assembly should be mounted and tested by the valve manufacturer. Since valves are "sources" of leaks of flammables, the limit switch may be in a classified location. Hermetically sealed contacts are highly desirable for limit switches. They should also be watertight NEMA 3 or 4.

Control Valves

A wide variety of modulating control valves with different types of actuators, inner works, and materials, both internal and external, is available to meet a given service. Proper design, selection, sizing, and characterization are essential to provide stable, reliable, and safe control loop design. Poor control valve design can cause cavitation (which can eventually destroy the valve and piping), corrosion, leaking, and high maintenance. Control valves should be engineered using the data developed within the industry. As with measurement elements, knowing flow rates and fluid characteristics, including vapor pressure, is essential. Knowing the pumping system characteristics is also necessary. Computer programs for cavitation sizing, and so on, are available. Control valves should not be sized based on a fixed pressure drop but based on the total pumping system friction drop. A rule of thumb is the control valve pressure drop at maximum flow is one-third or more of the total friction drop of the system. The control valve installation should be accessible for maintenance. Control valve bypasses should only be used when the dynamics of the process permit manual control by a hand-operated bypass valve. Control valves can be fail open or closed in the same manner as on–off valves.

Control Valves versus Adjustable-Speed Motor Drives

Flow control can be accomplished by controlling the speed of an electrical motor driving a pump or fan. This option has been popularized to reduce power consumption. The trade-off involves eliminating the power loss across the control valve in exchange for the added cost and complexity of the adjustable-speed drive power electronics. Most control valves are oversized and are operating at a significant turndown. The energy loss, which is proportional to the flow rate and pressure drop across the valve, could be saved by varying the pump speed instead of pressure throttling to achieve flow control. The energy saved and payout for the added investment in a drive system over a fixed-speed motor increases with the ratio of the pumping friction head to the lift head, the size of the system (drive system cost per horsepower decreases with size), flow turn down in percent, and value of saved electrical power in thousands of dollars per kilowatt. For instance, if we assume the control valve pressure drop is 40% of the friction head loss at design flow, any system larger than 10 horsepower can be justified for a 100% total friction system. If the system friction is only 25% of the total pumping head, a system larger than 50 horsepower is required for break even. These examples are based on a 75% flow turn down and electrical power savings valued at $1500 per kilowatt. In general, any drive over 50 horsepower has a good chance of covering the added cost in a short period of time.

Adjustable-speed drives can also have significant safety and environmental advantages if the process fluid is flammable, corrosive, toxic, or a pollutant by eliminating the control valves. Control valves leak and require maintenance. By eliminating the valve, leakage and the possibility of exposing maintenance personnel to toxic streams will be eliminated. In some cases, the process fluid may be so corrosive that a reliable control valve is not available.

Significant advances in adjustable-speed drive technology have resulted in significant improvements in performance and reliability. The new drive equipment, whether it is an adjustable-frequency ac motor drive or a dc motor drive, uses digital microprocessor-based controls that provide significant improvements in reliability and performance. They have digital communication capability and can talk to process computers or PLCs. They have improved diagnostics and can be programmed by a laptop computer or handheld programmer. There are also significant improvements in power semiconductors which make the drives more attractive. Power transistors are replacing SCRs, thereby eliminating complex commutation circuits required by SCRs.

ALTERNATING CURRENT VERSUS DIRECT CURRENT DRIVES

Alternating current drive electronics control the speed of standard ac motors by varying the supply voltage and frequency to the motor. It can be used to control the speed of an existing motor or be supplied as a new motor drive system. The ac drive electronics is larger and more complex than dc drive electronics, but the motor is standard (Figure 7-1).

Direct current drive electronics controls the voltage to the armature of a dc motor. Direct current motors are more complex than ac motors and require added maintenance of the armature brushes. In a Class I, Division 2 location, the dc motor would have to be explosionproof or pressurized because of the arcing brushes, whereas a standard induction motor without

FIGURE 7-1. Direct current motor drive system consisting of a dc motor with a cooling fan on the right and the drive speed controller on the left. The top of the drive controller includes a main breaker (note the warning sign concerning high voltage), control relay, and other power components. The bottom section has a removable cover plate that protects the printed circuit boards, capacitors, and other components. This cover plate is removed during the initial installation to provide accessibility to the terminal connections to connect cables from the controller to the motor and other devices. The dc motor on the right includes a cooling fan on the top and a terminal box on the left. (Reliance Electric Company, Cleveland, Ohio.)

contacts is acceptable. For this reason, ac drives may have an advantage over dc drives in chemical-process facilities.

CONCLUSION

The selection of the proper measurement and final control elements for process control systems is the key decision in designing control systems. A control system is only as safe and reliable as the measurement and control system.

There are a number of types of measurement devices, each with its area of application. It is always wise to use proven devices for the particular service.

There are a number of new devices that have significant advantages over traditional elements. Vortex and Coriolis mass flow meters are examples.

Adjustable-speed motor drives instead of modulating control valves should be considered, especially in sizes larger than 50 horsepower. They can save energy to justify their added cost, and they eliminate a leak source, the control valve.

REFERENCES

Hix, A. H. 1972. Safety and Instrumentation Systems. *Chemical Engineering Progress* 68 (5, May): 43–52.

Kletz, T. A. 1980. Don't Let Control Loop Blunders Destroy Your Plant. *Instruments and Control Systems.* (Nov.): 29–32.

Liptak, B. G. 1969. *Instrument Engineers' Handbook Volume I Process Measurement.* New York: Chilton Book Company.

Spitzer, D. W. 1984. *Industrial Flow Measurement: IRP Student Text.* ISA Publications Dept.

8

Process Control Safety

The results of the industry survey How Safe is Your Plant? in *Chemical Engineering Magazine* (April 11, 1988) indicate that accidents are sometimes attributable to improper operation of various process safety systems, including instrumentation and controls, equipment interlocks, emergency venting, and shut down systems. Most people involved in chemical-process facilities would agree with that position. Instrumentation systems that have failed and interlock systems that have misoperated have resulted in serious safety situations.

The proper design, installation, and maintenance of process control systems are essential to the safety of chemical-process facilities and the prevention of spills and environmental releases. Plants and companies can cite numerous examples of incidents that have occurred because of process control system failures: a styrene charge tank that overfilled because an interlock was bypassed, a batch sequencer that locked up and caused misoperation of control valves, and so on. Furthermore, a start up of a new process facility was harassed by "ghost" alarms caused by electrical noise. In another start up, a large motor burned out because of repeated cycling due to a software error.

Control systems must control the various parts of the process in such a way as to produce salable product to meet the commercial needs of the facility, but they must do so in a way that provides safe operation and eliminates environmental releases. Safety includes protection of equipment, facilities, and people, including operators who work in the process area, control room operators, maintenance people, people in the adjacent units, and neighbors outside the plant facility.

Control systems are designed to maintain the process within the required limits of temperature, pressure, flow, and level under normal conditions responding to the normal process variations, but the control system must also respond to the abnormal conditions that occur due to process upsets and take corrective actions to prevent dangerous conditions from occurring, including shutting the system down before a dangerous condition is reached. These protective control systems sense abnormal conditions and take actions through final control elements to prevent excess temperature, pressure, excess flow, overfilling vessels, excess vibration, or other conditions that can damage equipment or cause releases, spills, fires, or explosions from occurring. The abnormal conditions could be caused by equipment or piping failures, operational mistakes, or control system failures. No process can be totally free of failures or mistakes, considering the adverse environment and operating conditions present in chemical processes, but the system must be capable of coping with upsets and preventing serious, life-threatening situations from occurring.

The degree of protection should be commensurate with the possible consequences. Where highly toxic or flammable materials are processed, special protective measures should be used. The volume of hazardous waste and energy stored in these materials should be considered.

Process control systems consist of measuring elements and their transmitters, output final controls, and control room controllers (e.g., interlock relay panels, PLCs, microprocessors, miniature electronic controllers and recorders, annunciators, recorders and data loggers, distributed control systems, including local control units and operators' consoles, and process computers). These systems can be designed, installed, maintained, and operated to maintain process system safety and protection against environmental releases.

HISTORY OF PROCESS CONTROL

Figure 8-1 describes the evolution of industrial control technology from pneumatic controllers to distributed control. The changes in process control technology also include the transition from field operator control to the centralized control room and from hard-wired control components to software controlled microprocessors systems.

A process control room installed in the 1960s or 1970s typically consisted of a number of vertical operator's panels filled with miniature electronic and other controllers. The front of the control panels, facing the operator, included annunciators, recorders, individual controllers, selector switches, run lights, readouts, and devices that indicated the status of the process and provided the controls to operate the process. The inside of the panels was filled with electronic transducers, switches, relays, power supplies, and much wiring.

A great deal of design and drafting effort was required to install these systems, and changes required rewiring and recalibration of components. Each loop had to be checked out, and proper calibration and wiring had to be verified. Changes were difficult to manage and in some cases almost

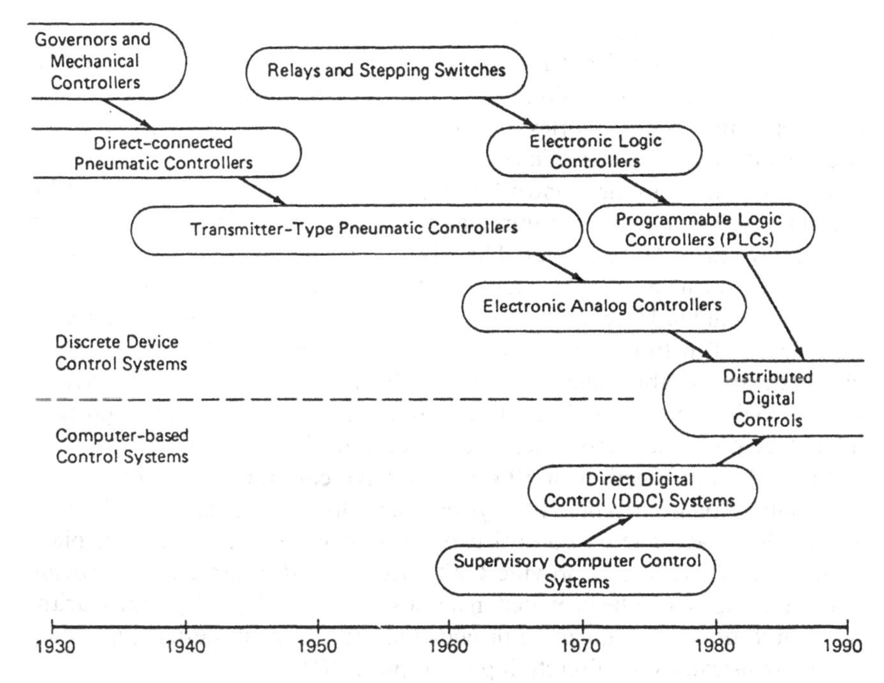

FIGURE 8-1. Evolution of Industrial Control Technology. (Lukas, M. P., *Distributed Control Systems, Their Evaluation and Design,* Van Nostrand Reinhold, 2, Fig. 1.1.)

impossible to verify because of wiring congestion. Electrical interlocking was accomplished with relays or if the process was batch, electromechanical drum programmers, that required extensive wiring between components.

Supervisory or direct digital control was used with large process computers that required a great deal of complex software effort. Distributed control systems provide some degree of success in providing high-level control, but the interaction with the hard-wired world was not easily or gracefully accomplished.

The introduction of the microprocessor and distributed control revolutionized control rooms, doing away with control panels and replacing them with operator's cathode ray tube (CRT) consoles. Wiring and discrete components have been replaced by microprocessors and software. The engineering effort in individual component and wiring design has been replaced and exceeded by process control software engineering. This process control revolution was driven by dramatic developments in components and hardware, software developments, and the industry need for improved process control to reduce material and energy costs. The transition from discrete component boards to integrated circuits and to large-scale integrated circuits was accompanied by significant improvements in process control software and programming. The conversion to microprocessor control, solid-state memories replacing core memories, and the proliferation of CRT displays also occurred during this dramatic conversion.

Programmable Logic Controllers appeared in the early 1970s, driven by the automotive industry's requirement for a cost-effective replacement for large relay panels (Figure 8-2). The advantages of PLCs were immediately obvious, and this market has continued to expand in size and complexity.

Programmable logic controllers can be easily programmed to perform interlocking functions using relay ladder logic, are flexible, can easily be changed, are reliable, and can provide diagnostics. Because it is easy to change logic using a CRT programming panel, flexibility is their greatest advantage over relay and other electromechanical equipment.

Since the introduction of PLCs, they have continually improved their capability. Some current PLC systems are difficult to distinguish from DCS. They have analog control capability, can communicate with plant computers or DCS, can provide CRT displays and graphics, can provide alarm functions, can be provided in a hot standby mode, can perform arithmetic and loop control, can be provided in fault-tolerant designs, and may have intrinsically safe discrete input/output (I/O).

Distributed control systems appeared on the process control scene in the mid 1970s and since then have assumed the dominant role in process scheme and replaced discrete component analog systems. There are a number of manufacturers of different lines of products, but regardless of the product, the system design is based on distributed digital microprocessor control.

DISTRIBUTED CONTROL SYSTEMS

A generalized description of a DCS is shown in Figure 8-3. The systems provided by individual manufacturers will differ from this generalized discussion. In general, they will have greater communication capability between the operator's control console and the high-level human interface indicated in Figure 8-4.

The local control unit interfaces directly to the process via closed-loop control. Interaction with the local control unit to perform tuning, adjust set points and control modes, and, in general, configure the loop can be accomplished at the low-level human interface. Data collection is accomplished through the digital I/O unit. Programming of the local control unit is accomplished using function blocks and various control languages. Various control languages are available as described in Chapter 3 of *Distributed Control Systems: Their Evaluation and Design* by Lukas.

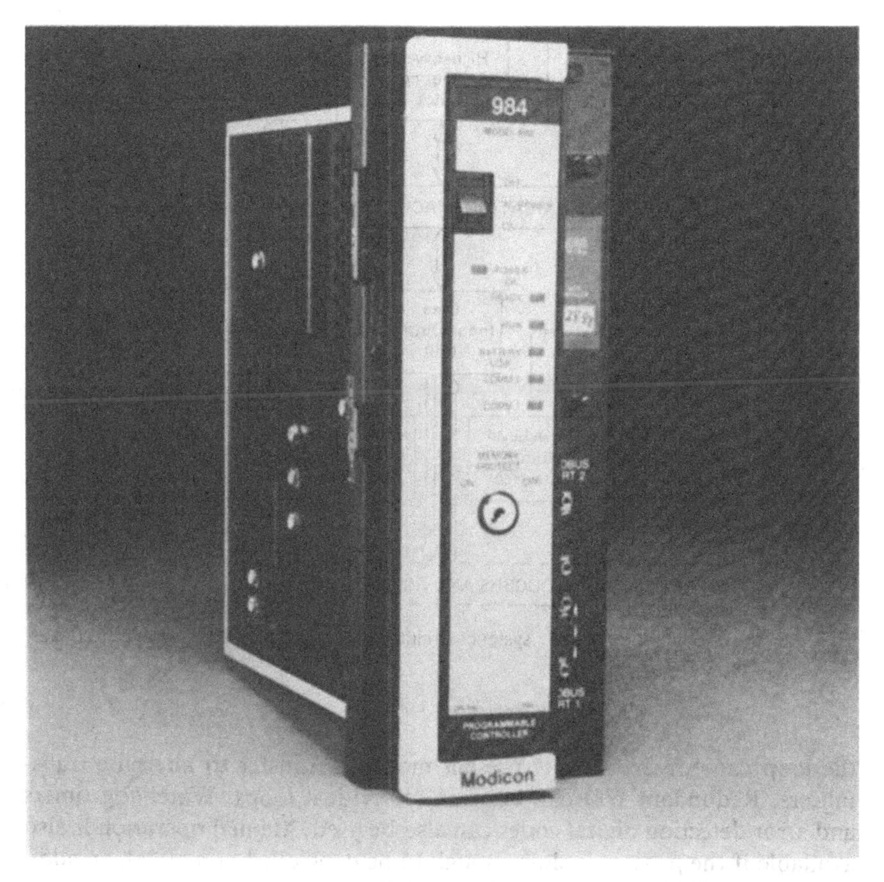

FIGURE 8–2. Programmable Logic Controller Modicon Model 984-680. A complete control system includes I/O modules, communications networks, and associated cables and option modules to suit individual applications. The slot mounted 984-680 is especially designed with high security features to protect against changes to the application logic program. These features are especially important in process control, boiler management, and other process control applications. (Modicon, Inc., Andover, Mass.)

Distributed control systems can be designed and configured to minimize the impact of hardware failures through diagnostics, redundancy, hot-standby systems, and manual controls. In evaluating the relative safety of a control system, it is prudent to evaluate these options. The first step in ensuring reliability is to use hardware and software systems that are proven and tested.

Distributed control systems can be designed to compensate for hardware failures. For instance, if a transmitter fails, the failure can be detected and

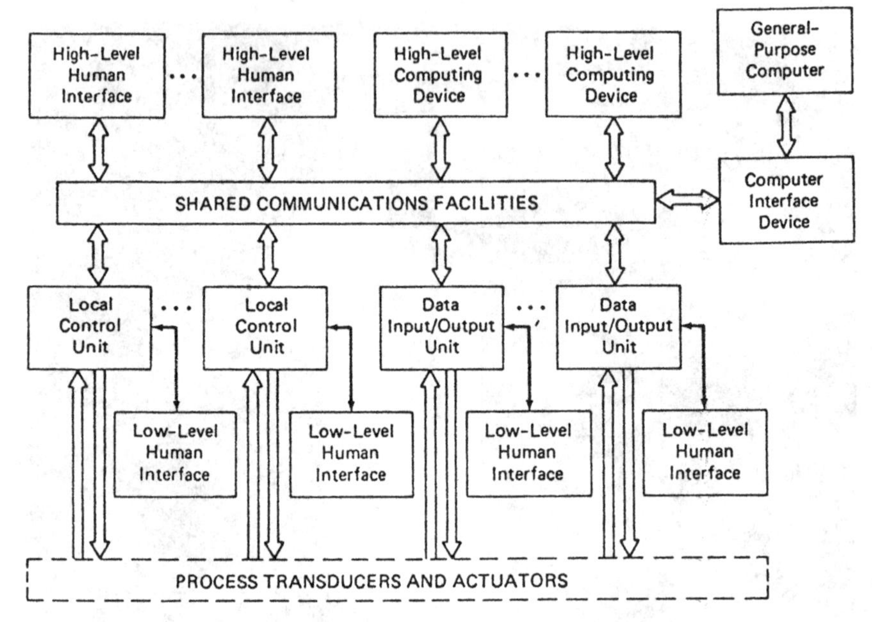

FIGURE 8-3. General control system architecture for a DCS system. (Lukas, M. P., 10, Fig. 1.4.)

the loop can "freeze" in its present mode or transfer to alternate transmitters. Redundant I/O can be used for critical loops. Watchdog timers and error detection digital codes can also be used. Manual operation is also available if the process is slow enough to be controlled in manual. In addition, the flexibility and versatility of DCS can provide many options to protect against software failures.

ALARM SYSTEMS

Alarms are indications of abnormal process conditions that require a response or action by the control room operator. Typical alarm conditions could be high pressure, high temperature, or pump failure. A typical control room using discrete miniature electronic panel instrumentation of the 1960s and 1970s would have many alarm or annunciator units, typically at the top of each panel. Alarms could be actuated by contact devices, for instance, a high-level switch, or by an analog controller. Annunciator system sequences are defined in ISA standard S181, *Annunciator Sequences and Specifications.* As an example, Sequence A with automatic reset consists of the steps shown in Table 8-1.

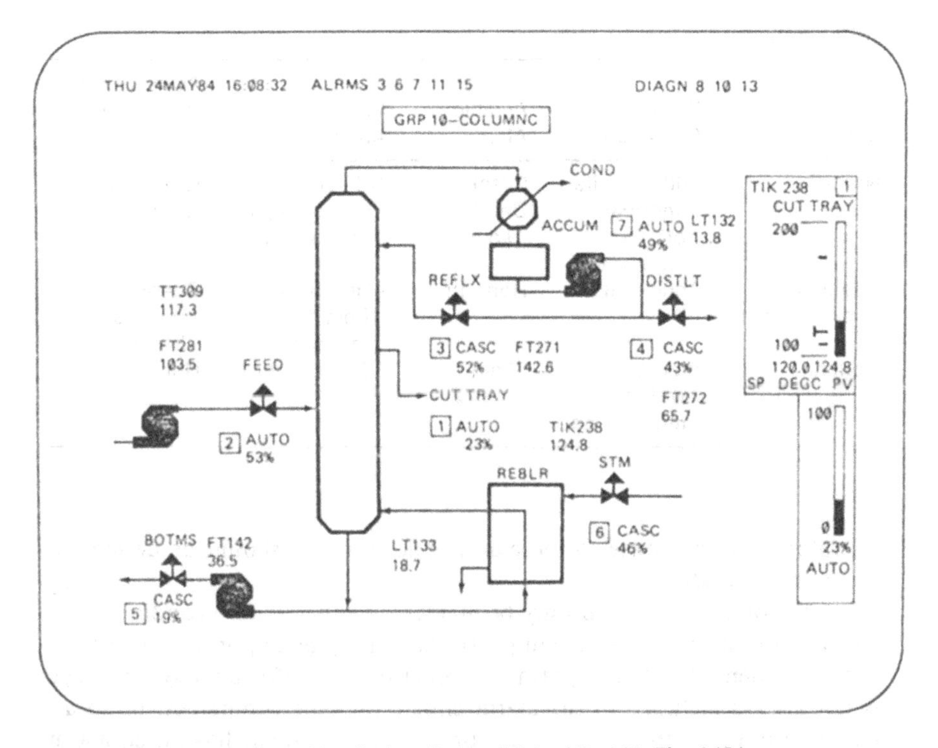

FIGURE 8-4. Control graphics display. (Lukas, Michael P., 197, Fig. 6.17.)

First out annunciator systems are used for shut down interlock systems, where one trip could cause a series of alarms to trip simultaneously. First out systems have the advantage of identifying the first initiating condition that caused the system to shut down. After the first out system trips and is acknowledged by the operator, the alarm that tripped first stays flashing while the other alarms in the first out sequence are indicated by a steady lighted window. This type of annunciator is essential for shut down interlock systems.

A common malady typical of many control rooms is overalarming. Seeing alarm windows flashing in a control room is typical. During even a mild upset, the control room operator may be overwhelmed with alarms. In some situations, the frequency of alarm may be so high that the system may have lost its credibility.

Most systems are overalarmed in an effort to inform the operator of every unusual occurrence. Control room operators, however, learn to identify critical situations. The important decision is what should or can the

TABLE 8-1.

Sequence State	Process Condition	Audible Alarm	Visual Alarm	Acknowledge Push Button Status
Normal	Within limits	Silent	Off—no light	Not pushed
Alarm	Abnormal off, too high or low	On—horn sounds	Flashing light on window	Not pushed
Operator acknowledge	Abnormal	Horn off	Light turns from flashing to steady	Operator pushes button
Normal	Process within limits or on	Silent	Off—no light	Not pushed

operator do in response to the alarm. The response should be defined in operating procedures.

Some conditions should only be alarmed if other failure conditions occur. For instance, a charge pump off alarm may be important during the charge sequence but not important at other times. Unfortunately, this logic is not always included in the alarm circuitry. Some companies, in an attempt to improve the effectiveness of the alarm system, have used alarm classification schemes of typically three classes. The highest and most critical class would be located and sequenced to command the operator's attention, the second class would be less critical and accordingly less prominent, and the lowest class (class 3) would be items that should receive some attention but not immediately (e.g., maintenance items that could be delayed). Class 3 alarms may be only status lights.

With the introduction and proliferation of DCSs, alarm system design and operation changed radically. Providing separate annunciator units and hardwire signals to the units is no longer necessary. Alarms now appear on process CRT screens on consoles as flashing lights, lists, or other indications. The entire process database in the DCS is available for alarms, and the alarm indication can flash on a process color graphics display (Figure 8-4). DCS provides capability to list, log, and prioritize alarms. The DCS system can analyze process conditions and suppress certain conditional alarms. It can determine if critical process conditions are occurring and indicate operator action. So called smart alarming is possible with DCS.

Some alarms are of such a critical safety nature, they should be hard wired directly to a separate annunciator unit. This should be done to provide separation from the normal process deviations and to protect these critical safety signals from possible software errors and hardware faults that can occur in a large DCS. There is also a class of common system alarms that is not a part of the process control system design that should be separated and isolated from the DCS design. In addition to critical process safety alarms, the separate hardwire annunciator module could include the following alarms:

Fire
Spill
Safety shower
Power system fault
Ground fault
Electrical and computer room smoke detector
Loss of ventilation
Other common system alarms

INTERLOCK SYSTEMS

Interlocks are on–off controls that provide sequence of shut down control of systems. They can control the start-up, operation, and shut down of systems. Interlocks are involved in the operation of most industrial processes.

In chemical processes, interlocks are involved in the operation of substation switch gear (key interlocks prevent the misoperation of power switches), in hazardous locations (to alarm or shut down if purging and pressurization systems fail per NFPA 496 or to protect explosionproof motors against excessive temperatures), and to prevent upset or unsafe conditions from occurring process interlocks are provided. This discussion is concerned with process control interlocks.

The design of the interlocks begins with the development of process and instrument flow diagrams. Process data, operations requirements, and equipment specifications are necessary to define the interlock requirements for a process. The design of the interlock is accomplished by the instrument, electrical, and process computer disciplines, but process design and plant operations and maintenance provide important inputs into the design process. Process design must ensure that the interlock design meets the process requirements, and plant operations must agree that the system meets operating requirements.

The interlock description must be defined in a way that is understandable by people not versed in electrical instrument systems. The basic requirements are usually defined by process design in general terms at the process and instrumentation (P and ID) diagram development and are expanded as the process design develops. The following methods of documenting interlock design can be used, depending on the type of process, company practices, and experience and preferences of the design group:

Word descriptors
Logic diagrams
Flowcharts
Programming matrices
Step descriptions (for batch processes)
Time sequence diagrams (for batch processes)
State transition diagrams

Simulation and final checkout are essential to ensure that the final control system installation meets the interlock description requirements. All interlock requirements must be documented to enable operations and the process control departments to understand them. Unfortunately, this is not always the case. As an example, a plastics sheet line process was experiencing mysterious interlock shut downs. The shut down investigation team asked to see the interlock description and discovered that the interlocks were not described anywhere except on electrical schematics. Batch control systems usually involve a significant amount of step sequence interlocking between the various material charging systems and various on–off valves, pump, and agitator motors. Continuous processes have fewer interlocks, but the system may be critical because of problems associated with shut downs and safety interlock schemes.

Manual controls are usually provided so the operator can control final elements (e.g., the pump or agitator motors or control valves) in the event the controller fails, or because of process equipment or piping failures, or because the process must be operated in an abnormal mode). Manual controls can be selector switches on a control panel or can be implemented in software by forcing a final control element on or off on a PLC monitor panel or DCS operator's console.

When a final control element is in manual, there are no interlocks except for critical hard-wired safety interlocks. An example of this situation is manual control of a charge valve to a batch reactor. This control could be hard wired to the batch reactor bottom valve open limit switch to prevent charging the material through the reactor, which could result in a serious process safety situation.

SAFETY INTERLOCK SYSTEMS

There is a growing concern in the chemical process industry regarding the design of safety interlocks. This has resulted from an industry concern, in fact, a commitment to preventing catastrophic releases and explosions, especially when toxic high hazard materials are processed, coupled with another concern with the application of PLCs and DCSs to critical safety interlock systems. The question is, can these systems be fail safe, considering the complexity of the microprocessor digital electronics and their vulnerability to software errors?

Committees in the ISA (SP 84) and AIChE are developing standards for the application of Programmable Electronics Systems, (PESs), to safety interlocks. Interlock systems can be classified according to the severity of the failure. The highest risk classification, Class I, involves situations where hazardous materials may affect people outside the plant boundary; Class II includes life-threatening situations within the plant; Class III involves major property damage within the plant. These classifications can be the basis for selecting appropriate protective measures.

For a given process situation, the first step in the design of safety interlock systems is to eliminate the need for the safety interlock by changing the process or equipment. This strategy is discussed in papers presented at the AIChE summer national meeting (August 19 and 20, 1990). Kletz, in Inherently Safe Plants—An Update, indicates that process design safety alternatives should be considered before adding protective systems. He cites some options, for instance, using less of the hazardous material, using it in a less volatile form, or not using it at all (i.e., substituting a less dangerous material).

Designing inherently safe systems should always be the first consideration; designing protective systems, particularly interlock systems, should then be considered. There are a number of alternatives and options in providing safety interlock systems. Each plant should consider the individual situation in light of its own environmental conditions and the risks involved. The following is a brief discussion of some of the alternatives and factors to be considered.

Interlocks are not a replacement for good process or mechanical or piping system design, installation, inspection or maintenance. Process containment is the key issue. No safety interlock can protect against premature rupture or failure of the process equipment or the piping system under normal process conditions. Only proven and tested components and installation methods should be considered for critical applications. Human factors, such as operator response to upset conditions and maintenance of critical interlock systems, should be considered. In general, keep the system as simple and straightforward as possible; avoid complex and intricate designs.

Relay hard-wired systems have been used extensively for interlocks and are generally considered fail safe. That is, relays fail to the deenergized condition. For mechanical relays, this is true if the coil opens or a lead wire breaks. The relay armature deenergizes and a normally open contact will shut down the process. There are, however, other modes of relay failures. If the spring that holds the armature open breaks, then, depending on the relay design, the normally open contact may not open. If the contact is fused by a short circuit, it may not open when the coil is deenergized. If water enters the relay enclosure and bridges between the normally open contact, the interlock circuit may not deenergize when the normally open contact opens. Finally, if the wiring to the normally open contact is long (typically over 1000 feet), the distributed capacitance of the wiring could provide enough leakage current in the ac systems to hold in the circuit even though the normally open shutdown contact has opened.

Fail safe for hard-wired systems using discrete electronic components means that the final element will fail in the safe direction in the event of a power failure or wire break, but it does not mean that the internal electronics will fail safe. Temperature electronic switches can be designed to fail upscale or downscale in the event the thermocouple burns out; in general, however, there is no guaranteed failure mode for many electronic modules.

The safety of any hard-wire system can be negated by interlock jumpers that bypass critical interlocks, and they may be difficult to find in the maze of control panel and control room wiring. Most plants have administrative controls to deal with interlock bypasses and generally require approvals of management. In some facilities, panel selector switches are used to bypass start-up interlocks, but when the bypass position is selected, a contact on the switch triggers an alarm.

PLC systems have evolved dramatically since the early versions. There were occasional failures, especially in output triacs, but generally the reliability was good. The early applications were primarily pure interlock systems that performed well. Some of the newer versions are sophisticated and resemble DCS systems. They can perform analog control and provide operator interface.

DCS systems have greater capability in the operator interface and sophisticated process control. PLC programming of interlocks is easy, whereas DCS systems use logic systems and tables that are more complicated to apply compared to electrical schematics. They may, however, appeal to chemical engineers. The PLC niche still appears to be primarily interlocking, whereas DCS is primarily regulator control and batch interlocking. Some systems consist of a DCSs, and PLCs in a coordinated regulatory and interlock control scheme.

Recent versions of PLCs have been developed to meet the demands of high reliability critical safety interlocks. Specialized interlock controllers called fault tolerant controllers have also been developed to meet these critical safety interlock requirements. These types of controllers are designed so that no single fault can cause a failure; in addition, they feature redundant logic systems (usually triplicate), two out of three voting logic, separation and duplication of functions, software testing (using internal and external watchdog timers), redundant output triacs, and redundant input processing. This type of equipment is designed to reduce the probability of controller error to essentially zero, but one should always carefully evaluate common mode failures, for instance, power supply failures.

Maintaining software security is also important. Safety interlock programming must also be protected against unauthorized changes.

DCS systems can be programmed to provide redundancy, but they do not in general have the voting logic or triple redundant features found in fault tolerant controllers or certain PLCs. Software security is a much greater issue in DCSs because of the size of these systems. Safety interlocks should be dedicated to fault-tolerant controllers or dedicated PLCs designed with specialized safety features.

ELECTRICAL PROTECTION OF CONTROL SYSTEMS

Noise, Surge Protection, Grounding, and Circuit Protection

The great interest in fault tolerant controllers, safety PLCs, and the reliability and failure modes of digital microprocessors would indicate that controller reliability and failure modes are a major problem. Just the opposite is true, however. The controller is the most reliable part of the system. I/O failures caused by primary element, final control element, or transmitter failures or operational problems caused by electrical noise, voltage surges, improper control system grounding, or control power circuit and component protection are far more significant than controller failures. One PLC manufacturer estimates that 80% of the failures occur outside the PLC.

Failures in input and output signals as a result of electrical noise, voltage transients, improper control system grounding, and control power circuit and component protection are major sources of control system safety problems. Input and output signals are the weakest link in the control system. In general, there is a failure to appreciate these problems. Proper wiring methods are essential to control system reliability and safety. As an example, Possible Hazard in PC Versatility in *Electrical Construction and Maintenance* (September 1984) described a facility where a number of push buttons were wired normally open instead of normally closed (the correct failure mode) and were programmed normally closed.

Protection of signal wiring systems by separation, the use of shielded and twisted pair cables, overvoltage protection, and circuit fusing are essential to reliable and safe control system operations. Examples of problems are numerous. In a recent start up, a dc power supply shortcircuited, but instead of opening the fuse that fed the supply, it caused the main control power breaker to open, shutting down the entire control system. Electrical noise problems can cause disruption of a control system and these can cause system shut downs.

Electrical noise is an electrical disturbance injected into control system wiring or components that can cause the control system to misoperate. Noise distorts the signal transmission. It can interfere with millivolt signals generated by thermocouples or load cells, milliampere signals from transmitters, digital pulses from a transmitter, microprocessor or computer, and 120-volt signals. The amount of interference depends on the frequency and magnitude of the noise source relative to normal signal level and the sensitivity of the control electronics. Low-magnitude signals like thermocouple or load cell signals are vulnerable, digital signals are sensitive to fast noise pulses, and even higher level signals are vulnerable if the signal electronics is sensitive enough.

Noise can be radiated into control system circuits and equipment by radio transmitters, particularly handie talkies. An operator in a control room was able to trip high-temperature alarms by operating his handie-talkie radio transmitter. Metallic enclosures shield electronic and microprocessor components from radio frequency interference (RFI), but RFI can also be picked up by the signal wiring, which can act as an antenna. If the control system is subject to RFI, the equipment must include RFI filtering or be desensitized to RFI and comply with industry RFI standards.

Electromagnetic interference can also generate voltages in signal wiring or enclosures via the magnetic field generated around any conductor. The magnitude of the field varies directly with the magnitude of the current and inversely with the square of the distance. If the current is ac, the field is ac. If the current wave shape contains high-frequency harmonics, for instance, the current to a motor drive, the induced field will include them. High-current surges from "inductive kicks" any time a starter coil or solenoid is deenergized can induce voltage spikes into control wiring. Twisted pair wiring is used to cancel out any magnetic field resulting from this source of EMI.

Noise transient voltages can also interfere with control signals by capacitive coupling of voltage changes into signal wiring and components. This coupling depends on the distance from the noise source and the magnitude of the voltage. Shielded signal cable is used to minimize this electrostatic noise coupling.

Electrical noise can also be injected into control system wiring by ground loops. Anytime a signal wiring system is connected to ground at more than one point, the ground system is part of the signal return path and any noise generated in the ground path, which usually includes building steel, is part of the signal system and can disrupt the signal circuit.

Electrical noise can also be injected into the control system by power supply disturbances that can be coupled through power supply transformers into signal wiring. The signal wiring system needs to be designed and engineered to minimize noise disturbances. Separation of signal wiring from power wiring is standard practice to prevent noise problems. The actual separation required depends on many factors and should be developed in consultation with electronic equipment manufacturers. Shielded and twisted pair instrument cables are used for signal wiring. Three-conductor power cables are typically used for power wiring. For motor and other balanced three-phase loads, the magnetic field around three conductor cables is zero.

Instrumentation signal wiring should be separated from power wiring by the use of separate conduit or cable tray systems or by providing a cable tray divider.

Signal and signal return conductors should always be in the same cable. If the signal return conductor is separate from the signal conductor, the area between the conductors is a loop for EMI pickup. When twisted pair or shielded cables are connected to equipment, the shield grounding should be connected as close to the connection as possible to minimize the length of conductors that are unshielded and not protected.

Overvoltages can cause instrument transmitters or other electronic microprocessor equipment to fail. One of the major sources of overvoltage is a lightning strike. It is not necessary for lightning to strike directly on or near the instrument. As the lightning discharge travels to earth, significant overvoltages can occur within a structure and disseminate to various parts of the structure. It is not unusual to find transmitters, especially those located on the top of a structure, that have failed after a lightning storm. Overhead cabling that is not shielded from a lightning discharge is especially vulnerable.

Transient voltages superimposed on the normal circuit voltage can be coupled or induced into signal wiring by the switching of the capacitive or inductive elements in the wiring system and the releasing of the stored energy. The switching can be the normal circuit contact action or intermittent wiring failures. The induced overvoltage spikes are only limited by the stored energy in the elements and the breakdown voltage of the wiring system. Electronic components, especially microprocessor chips with their very small voltage clearances, are vulnerable to voltage spikes. The failure may be immediate or may occur after a time.

Protecting input signal and power wiring may be necessary since lightning or switching overvoltages can find devious paths into systems. Radio hams are well aware of the damage that can occur when an aerial cable is exposed to a lightning discharge. Thus, surge protection devices can be connected at the input wiring to devices or systems.

There are a number of types of voltage surge protection devices that can be used to protect electronic devices, including metal oxide varistors (MOVs), zener diode-type devices, or gas tube discharge devices. The breakdown voltage of the device must be higher than the circuit operating voltage but lower than the level that will damage the protected component. The speed of response is important to provide protection against fast rise-time pulses, and the leakage current at normal operating conditions can affect the accuracy of the measurement. These devices also have energy limitations on the pulse that they must clip.

Metal oxide varistors are available in low-voltage dc, but they are susceptible to failure and blowing apart when subject to sustained overvoltage. In one instance, MOVs failed due to harmonic overvoltages and caused a fire in a dc motor drive cabinet.

The application of surge protective devices requires considerable judgment and experience and should be coordinated with the manufacturer of the components or electronic equipment to be protected. Protection against inductive kicks from relay, starter, or solenoid coils is a much simpler matter and can be accomplished with MOVs for 120-volt circuits and reverse diodes for dc circuits.

Data highway cables can also be protected against lightning surge voltages by inline surge voltage suppressors. Typically, gas tube protectors are used for fast transient response, but other types of data line protectors are available.

Instrument grounding is primarily an electrical noise issue. Instrumentation circuits are typically at 24 volts dc or less, and grounding for personnel protection is not required by the *NEC*. A 120-volt ac signal or power circuit must be grounded per Article 250 of the *NEC*. An equipment grounding system consisting of cable tray, conduit, and device, must be bonded together back to the supply transformer, and the supply transformer neutral must be connected to the equipment grounding system and to the system ground, which is usually building steel. The green insulated wire, or grounding conductor, can be used in lieu of conduit and cable tray as the equipment grounding conductor for solenoids, limit switches, and other 120-volt devices.

It is essential that limit switches, solenoids, and other 120-volt devices have their enclosures bonded to the equipment grounding system. This bonding connection is sometimes missed because the limit switch is connected to process piping, but the piping system is not an acceptable equipment grounding system by itself. For instance, if the limit switch were removed from the valve by maintenance personnel and the hot wire faulted to the metallic enclosure, a person contacting the limit switch enclosure could be shocked.

Instrumentation signal wiring systems should have the wiring system grounded at one and only one point—the signal source. From 4 to 20 milliampere loops are grounded in the control room, while thermocouples should be tip grounded. Ungrounded instrument wiring invites noise problems and voltage transients, but the effectiveness of the grounded system depends on maintaining single point grounding. Ground loops are more than one connection of wiring system to building steel or earth, and any noise voltages that occur in building steel or earth are coupled into the signal circuit. Electrostatic shields should be grounded only at one point, preferably at the wiring system ground.

In wiring systems, ground loops occur and may be difficult to find and correct. An electronic device called a coupler or dc input isolating conditioner can be used to open the ground loop. It can be matched to the loop and provide wiring isolation. In a sense, it acts like a dc transformer and can be used to solve ground loop noise problems.

Computer grounding has been a controversial issue because some of the recommendations of some computer manufacturers are at odds with some of the principles of electrical equipment grounding. The isolated ground specified by some manufacturers is a separate grounding system connected to its own, separate low-resistance grounding electrode. The intention of this system is to avoid noise interference generated by the electrical stray noise currents that may be present in building steel or the earth grounding the system. Isolated systems invite problems because the systems are never truly isolated. There is always coupling by connection to the same earth or building plane and capacitive coupling through the system wiring. Isolated systems are not truly isolated any more than ungrounded systems are truly ungrounded. Isolated systems can be a shock hazard if the equipment grounding system is missing.

The best ground system, whether it is static electrification, electrical system and equipment protection, or noise elimination, is the one that bonds all metal parts together and connects the wiring system to building steel or earth at one point and only at one point. The culprit is voltage differences within the system. Electrical equipment and system grounding as described in Article 250 of the *NEC* prevent shock voltages and provide a low-impedance path through the equipment ground conductors, whether they are a green ground wire, conduits, cable trays, bonding jumpers, and so on, for any fault current back to the supply transformer to achieve operation of the breaker or fuse.

Electrical equipment grounding of separately derived systems (a separately derived system is established whenever a transformer, generator or UPS system is provided) requires that the neutral, or grounded conductor is connected to a building steel or earth at one and only one point. This is referred to as single point grounding. If the neutral is connected to building steel or earth at more than one point, either intentionally or by accident, a power ground loop is established and stray currents can flow in building steel or the earth ground. Poor power grounding can contribute to computer noise problems.

Instrument systems or equipment that use a chassis or panel frame as part of the signal return wiring cause ground loop noise problems.

Extensive ground rod installations and chemical grounds provide good electrical contact with earth, but bonding all metallic enclosures together is far more important. A zero-impedance ground plane provides the most effective grounding system for all grounding problems. If for various reasons stray currents cause interference with computers, verify that all metallic enclosures and parts are bonded together, check that the electrical power system is grounded correctly and that neutrals are not connected to building steel except at one point, provide an isolation transformer as close as possible to the computer equipment, and reduce the impedance of the ground plane. The Ufer ground system, connection of concrete rebar together in a ground grid, can be an effective ground grid. Where building steel members are corroded or insulated by painted members, bonding jumpers between various members may be desirable to reduce the ground system impedance.

Where high-frequency interference is a concern, the length of the equipment ground conductors can be a significant factor, and a multipoint equipment grounding grid system has been proposed. At very high frequencies, the conductor can produce reflections if the conductor length is greater than $\frac{1}{20}$ the wavelength of concern. The multipoint grounding system, however, has a greater sensitivity to low-frequency interference. A sensitivity to high-frequency interference must be demonstrated to justify this approach.

Control system circuit protection consists of the fuses, circuit breakers, and power distribution panels that provide power (120, 240, or 208 volt ac or low-voltage dc) to computers, PLCs, special transmitters that require separate fusing, and other process control systems. One of the mistakes typically made in the design of these systems is the installation of multiple process systems on the same fuse. In one situation during tie ins to process system A, a failure occurred that blew the control power fuse to A as well as shut down lines B, C, and D that had their control systems powered from the same fuse.

The control power distribution system should be designed to provide individual fusing to equipment and process systems. When a control circuit fuse blows, the effect on the operation of the process control system should be minimized. An entire line, reactor, or building should not be shut down if a control power fuse opens. For that reason, PLC fusing of inputs, outputs, and processor has some distinct advantages: If an output fails, the system still operates. It is much easier to isolate the failed component. Individual fusing provides better protection for components. The current rating of the fuse can be closer to the component current rating, so when a failure occurs, the damage to the component is less.

The type and rating of the fuse should be coordinated with the load characteristics to provide maximum protection with opening under normal load conditions. Branch circuit fuses that protect individual components should be selected and coordinated so if a failure in a component occurs, the main power fuse does not open. Control power for smaller systems should be supplied by 120-volt systems, not 240- or 208-volt systems unless larger load requirements require a higher voltage. Systems of 120/240 or 120/208 volts can have neutral wiring failures that can cause overvoltages to occur. During one start-up, the neutral from a 240-volt three-wire transformer opened and the load distribution was such that the voltage to the instrumentation recorders rose from 120 to 180 volts, causing the instrumentation to fail.

Finally, electrical noise-generating loads, such as welding outlets, adjustable speed drives, and fluorescent lights, should not be on the same transformer secondary as sensitive microprocessor loads that can malfunction due to conducted electrical noise. Welding in the vicinity of computer equipment or other controls that use microprocessor electronics can be damaging to the electronics if the welding current returns through building steel.

CONCLUSION

Control system reliability is an important safety consideration in chemical processes. Control system failures have caused serious safety incidents. Distributed Control Systems provide improved reliability over discrete miniature electronic systems. They can also be programmed to provide redundancy. They are well adapted to continuous control systems but are less adapted to interlock systems. Programmable Logic Controllers are very effective in performing interlock control but are not as applicable to continuous process control.

Where critical safety interlocks especially Class I interlocks, are involved, they should not be included in PLC or DCS systems unless special controls are used.

Fault tolerant PLCs provide an option for critical interlock requirements. In addition, redundant measurement and control elements should be considered. Software security and redundant hardware are important issues.

Control systems should be protected against overvoltages, electrical noise, and component failures. This can be accomplished by proper grounding, separation of instrument wiring from power wiring, use of transient voltage suppressors, and use of shielded and twisted pair instrument cables.

Improper grounding by providing isolated grounding systems for computer and instrument systems can violate the *NEC* and cause noise and overvoltage problems.

The principles of power and instrument grounding are similar.

REFERENCES

Buschart, R. J. Computer Grounding and the National Electrical Code. IEEE Paper PCIC-85-8C.

Cook, R. W. 1985. Suppression of Transient Overvoltages on Instrumentation Wiring Systems. *IEEE Transactions on Industry Applications* IA-21 (6, Nov./Dec.).

Freund, A., ed. 1987. Protecting Computers From Transients. *Electrical Construction and Maintenance Magazine* (Apr.).

Hix, A. H. 1972. Safety and Instrumentation Systems. *Chemical Engineering Progress* 68 (5, May).

Institute of Electrical and Electronics Engineers. 1982. IEEE Standard 518. IEEE Guide for the Installation of Electrical Equipment to Minimize Electrical Noise Inputs to Controllers.

John Barnard Co. 1985. State Transition Diagrams: Specifying Programmable Controller Logic without Relay Ladders. *In Tech Magazine* (Oct.):51–53.

Kerlin, T. W., and R. L. Shepard. 1982. Industrial Temperature Measurement. ISA Instructional Resource Package. Student Text.

Kletz, T. A. 1980. Don't Let Control Loop Blunders Destroy Your Plant. *Instruments and Control Systems* (Nov.):29–32.

Klipec, B. E. 1967. Reducing Electrical Noise in Instrument Circuits. *IEEE Transactions on Industry and General Applications* IGA (Mar./Apr.).

Krigman, A., ed. 1985. Relay Ladder Diagrams: We Love Them; We Love Them Not. *In Tech Magazine* (Oct.):39–44.

Lee, R. H. Grounding of Computers and Other Similar Sensitive Equipment. IEEE Paper PCIC-85-8B.

Lewis, W. H. 1985. Recommended Power and Signal Grounding for Control and Computer Rooms. *IEEE Transactions on Industry Applications* IA-21 (6, Nov./Dec.).

Liptak, B. G., ed. 1969. *Instrument Engineers Handbook. Vol 1 Process Measurement.* New York: Chilton Book Co.

Lukas, M. P. 1986. *Distributed Control Systems: Their Evaluation and Design.* New York: Van Nostrand Reinhold.

Musterer, R. 1989. A Case for Programmable Controllers in Process Control. *Inst and Control Systems* (Jun.).

Rosenof, H. P. and A. Ghosh. 1987. *Batch Process Automation: Theory and Practice.* New York: Van Nostrand Reinhold.

Scientific Apparatus Manufacturers Association. 1978. SAMA Standard PMC 33.1, *Electromagnetic Susceptibility of Process Control Instrumentation.*

Shaw, J. 1985. Smart Alarm Systems: Where Do We Go From Here? *In Tech Magazine* (Dec.).

Spitzer, D. W. 1984. *Industrial Flow Measurement.* Instructional Resource Package. Student Text. Instrument Society of America.

Wott, H. 1976. *Noise Reduction Techniques in Electrical Systems.* New York: John Wiley and Sons.

9

Electrical and Process Control Safety Standards for Chemical Processes

Standards and Codes . . . the Key to Safe Plant Operations, Quality and Profitability is the title of the Engineering Newsletter of January 1990 issued by the CMA. The article indicates that the CMA's involvement in the consensus standards process is aimed at promoting safety and quality and at enhancing profitability.

National standards and codes provide a valuable resource to those involved in the design, installation, maintenance, and management of chemical-process facilities. They represent the expertise of those who have experience in specialized areas. They are developed, reviewed, voted on, and approved by committees of individuals representing various interests and thus represent a consensus as to recommended practice.

National standards are dynamic in that they are updated as new technology, experience, and data become available; American National Standards must be updated at least every 5 years. National standards are developed by a voluntary program; people who work on the committees are sponsored by their company or organization. In addition, standards organizations have staffs to facilitate and coordinate the work of the committees.

The development or revision of concensus standards requires a public review. The American National Standards Institute, recognizes most standards writing organizations as Accredited Organizations that have established procedures for obtaining public review. Negative votes or comments must be reconciled. The new or revised standard is publicized in appropriate publications and in the ANSI newsletter to obtain further public review. This process ensures that all interested parties have the opportunity to comment.

NATIONAL STANDARDS ORGANIZATIONS

The following organizations are responsible for national standards related to electrical and process control safety:

The National Fire Protection Association (NFPA)
1 Battery March Park
P. O. Box 9101
Quincy, MA 02260-9101
Tel. No. 1-800-344-3555

Instrument Society of America (ISA)
67 Alexander Dr.
P. O. Box 12277
Research Triangle Park, NC 27709
Tel. No. 919-549-8411

The Institute of Electrical and Electronics Engineers, Inc. (IEEE)
The Standards Department
445 Hoes Lane
P. O. Box 1331
Piccataway, NJ 08855-1331
Tel. No. 1-800-678-IEEE

American Petroleum Institute (API)
1220 L. Street NW
Washington, DC 20005
Tel. No. 202-682-8000

Canadian Standards Association (CSA)
178 Rexdale Boulevard
Rexdale Ontario, Canada
M9W1R3

National Electrical Manufacturers Association (NEMA)
2101 L. Street NW
Washington, DC 20037
Underwriters Laboratories (UL)
333 Pfringsten Road
Northbrook, IL 60062
Tel. No. 312-272-8000

United States Department of Labor
Occupational Safety and Health Administration (OSHA)
200 Constitution Avenue NW
Washington, DC 20210

International Standards
International Electrotechnical Commission (IEC)
1 Rue de Varembe
1211 Geneve 20 Switzerland

National Fire Protection Association

The NFPA is an independent, nonprofit organization whose purpose is to promote fire protection and establish safeguards against loss of life and property by fire through the development of voluntary consensus codes and standards. It is the most widely recognized fire protection organization in the world.

The basic philosophy of the NFPA is that compliance with rules requires voluntary acceptance and participation of interested and affected parties. The NFPA system is based on voluntary participation, procedural fairness, and cooperation between governmental and private sector groups. The NEC is the prominent standard developed by the NFPA. For many years, the *NEC* has been approved for legal adoption by most state and municipal governments as well as by agencies of the federal government.

Forty-two states, as well as cities, towns, and counties in all 50 states, have electrical laws based on the *NEC*. The *NEC* is also the basis for OSHA regulations.

The present structure on the *NEC* consists of a main correlating committee and 20 code-making panels, each covering certain sections of the code. NFPA committees, including code-making panels, operate under the NFPA Regulations Governing Committee Projects. Each code panel consists of people and groups with diverse interests, including electrical contractors, inspectors, manufacturers, testing labs, insurance organizations, organized labor, governmental agencies, and private and public users of electricity. No more than one-third of the panel can be members of any single group.

The *NEC* is reissued every 3 years after a rigorous development cycle in which proposed changes are considered. The first step in the development cycle is a call for proposals from the public for changes to amend the current code. The appropriate code panels meet to review these proposals and act on each one. The results of the code-panel actions are published in the *Technical Committee Report* (*TCR*) for public review. Public comments on *TCR* are reviewed and voted on by the respective code-making panels, and the results are published for public review in a second document called the *Technical Committee Documentation* (*TCD*). The *TCD,* along with a preprint of the proposed *NEC,* is presented at the NFPA Annual Meeting for adoption by the membership.

The *NEC* is not a design specification nor is it an instruction manual for untrained personnel. The requirements in the *NEC* are minimal to protect people and property from the hazards that occur from the use of electricity. The code is not aimed at continuity of service or efficiency but at primary safety.

NEC articles of greatest interest to chemical process safety are the following:

500, 501, 502 Hazardous Locations
110 Requirements for Electrical Installations
240 Overcurrent Protection
250 Overcurrent Protection
300 Wiring Methods
318 Cable Trays
427 Fixed Electrical Heating Equipment for Pipelines and Vessels
430 Motors, Motor Circuits, and Controller's
450 Transformers and Transformer Vaults
685 Integrated Electrical Systems
702 Optional Standby Systems
725 Class 1, Class 2, and Class 3 Remote Control, Signaling, and Power-Limited Circuits

Other NFPA standards of interest for chemical process safety are as follows:

NFPA 30 Flammable and Combustible Liquids Code
NFPA 69 Explosion Prevention Systems
NFPA 70B Electrical Equipment Maintenance
NFPA 70E Electrical Safety Requirements for Employee Workplaces
NFPA 77 Static Electricity
NFPA 78 Lightning Protection
NFPA 325M Fire Hazard Properties of Flammable Liquids, Gases and Volatile Solids
NFPA 496 Purged and Pressurized Enclosures for Electrical Equipment
NFPA 497A Classification of Class 1 Hazardous Locations for Electrical Installations in Chemical Plants
NFPA 497M Classification of Gases, Vapors and Dusts for Electrical Equipment in Hazardous (Classified) Locations

Instrument Society of America

Standard and Practices Committee, SP 12, Electrical Equipment in Hazardous Locations, has led the way in the development of standards for hazardous locations. Since the early 1950s, this committee has pioneered in the development of such standards for chemical-process facilities. In the early days, it developed and promoted intrinsic safety in the United States. For example, the papers presented at the ISA Wilmington, Delaware, conference in 1964 are classics in their approach to safety. The topics included Intrinsic Safety, Area Classification, Purging, Explosionproof, Sealing, Dusts, Explosion Fundamentals, Fusing, Human Safety, Group Classification of Chemicals, Static Electricity, and German Standards. The SP 12 committee also developed standards on Purging, Intrinsic Safety, Nonincendive Equipment, and Dust Safety. The dust safety standard was the only one on dust classification until NFPA 497B was developed and issued in 1989. X, Y, and Z purging began in this committee.

The SP 12 committee consists of a main committee and subcommittees assigned to particular standards. Many members of the committee are also members of the *NEC* panel 14 and IEC TC31, Electrical Apparatus for Explosive Atmospheres.

The standards in which the SP committee is still involved include the following:

1. ANSI/ISA S12.10-1988 Area Classification in Hazardous (Classified) Dust Locations
2. ANSI/ISA RP12.6 Recommended Practice—Installation of Intrinsically Safe Systems for Hazardous (Classified) Locations, November 30, 1987
3. ANSI/ISA S12.12 Electrical for Use in Class I, Division 2 Hazardous (Classified) Locations
4. ANSI/ISA 12.13 Performance Requirements—Combustible Gas Detectors, Part I, 1986
5. ANSI/ISA 12.15 Performance Requirements for Hydrogen Sulfide Detection Instruments (10-100 PPM) Part I, 1990
6. ANSI/ISA 12.15 Recommended Practice—Installation, Operation and Maintenance of Hydrogen Sulfide Detection Instruments, Part II, 1990

Work is also underway on standard terminology, definitions for classified locations, and increased safety, an IEC concept.

ISA safety standards outside of the SP 12 hazardous (classified) locations area include

ISA ds 84.01 Draft Standard, Programmable Electronic Systems

The work of this SP 84 also includes subcommittees covering

SP 84.02 Reliability Calculations
SP 84.03 Hardware and Software Failures

SP 84.04 Applications/Operations/Maintenance

Other ISA safety standards include the following:

ANSI/ISA Safety Standard for Electrical and Electronic Test, Measuring, Controlling, and Related Equipment
- 82.01 General Requirements
- 82.02 Electrical and Electronic Test and Measuring Equipment
- 82.03 Electrical and Electronic Process Measurement and Control Equipment

These standards cover protection from electric shock, fault conditions, explosion, and fire for instrumentation.

Institute of Electrical and Electronic Engineers

The IEEE has an extensive standards program covering all phases of electrical technology. It is the largest technical organization in the world. The more relevant standards of safety in chemical process facilities follow:

- 141-1986 (IEEE/ANSI) Recommended Practice for Electrical Power Distribution for Industrial Plants (IEEE Red Book)
- 142-1982 (IEEE/ANSI) Recommended Practice for Grounding of Industrial and Commercial Power Systems (IEEE Green Book)
- 242-1986 (IEEE/ANSI) Recommended Practice for Protection and Coordination of Industrial and Commercial Power Systems (IEEE Buff Book)
- 303-1984 (IEEE/ANSI) Recommended Practice for Auxiliary Devices for Motors in Class 1—Groups A, B, C, and D, Division 2 Locations
- 446-1987 (IEEE/ANSI) Recommended Practice for Emergency and Standby Power Systems for Industrial and Commercial Applications (IEEE Orange Book)
- 576-1989 (IEEE/ANSI) Recommended Practice for Installation, Termination and Testing of Insulated Power Cable as Used in the Petroleum and Chemical Industry

American Petroleum Institute

The API has a standards program directed at the petroleum industry. It has developed standards relating to piping, vessels, and other topics. Its electrical and process control safety standards include the following:

- Classification of Locations for Electrical Installations in Petroleum Refineries, API Recommended Practice 500A, Fourth Edition, January 1982
- Recommended Practice for Classification of Locations for Electrical Installations at Drilling Rigs and Production Facilities on Land and on Marine Fixed and Mobile Platforms, API Recommended Practice 500B, Third Edition, Oct 1, 1987
- Classification of Locations for Electrical Installations at Pipeline Transport Facilities, API Recommended Practice 500C, Second Edition, July 1984
- Recommended Practice for Electrical Installations in Petroleum Processing Plants, API Recommended Practice 540
- Recommended Practice for Design and Installation for Electrical Systems for Offshore Production Platforms, API RP14F, Second Edition, July 1, 1985
- Protection Against Ignitions Arising out of Static, Lightning and Stray Currents, API Recommended Practice 2003, Fourth Edition, March 1982

Canadian Standards Association

The *Canadian Electrical Code* (*CEC*) and system of standards and approval agencies are similar to the *NEC*'s The *CEC* is different from the *NEC* in some important ways. The *CEC* consists of the 1990 *CEC* Part I, Sixteenth Edition C22.1-1990, Safety Standard for Electrical Installations. The topics in this code are similar to those in the *NEC:* general rules, conductors, services, grounding and bonding, wiring, Class 1 and 2 circuits, hazardous locations, and so on. Hazardous (classified) locations are defined in a similar manner to Article 500. The *CEC* appendixes include Safety Standards for Electrical Equipment, which provides a listing of standards used by the CSA to test and certify electrical equipment. In a sense this is similar to U.S. UL listing of safety standards. The appendixes also include sections on installing intrinsically safe and nonincendive electrical equipment and wiring and on using combustible gas detection instruments in Class I, hazardous locations.

The *CEC* permits the use of combustible gas detectors in situations where listed equipment for Division 1 or Division 2 locations is not available. If the classified location is Division 1, Division 2 equipment can be used; if the classified location is Division 2, general-purpose equipment can be used. The combustible gas detectors must be applied in conformance with the indicated standards. The *CEC* also permits cables conforming to CSA standards in Division 1 locations. Neither of the options is permitted by the *NEC*.

The *CEC* is administered by a main committee and subcommittees; facilities, which vary for each province, are inspected by local authorities.

An excellent reference on Canadian standards is *Hazardous Locations* by John Bossert and Randolph Hurst available from the Canadian Standards Association.

National Electrical Manufacturers Association

National Electrical Manufactures Association is the largest trade association of electrical products in the country and a leading organization in the development of voluntary standards. NEMA standards express an agreement among manufacturers that they will meet the criteria in the standards. The standards cover all varieties of electrical products, including circuit breakers, conduit, cables, ground fault circuit interrupters, panel boards, and switch gear NEMA also provides informational standards relating to electrical products.

NEMA standards that apply most directly to electrical and process control safety include

NEMA VE-1-1984 Metallic Cable Tray Systems
NEMA 250-1985 Enclosures for Electrical Equipment (1000 volts maximum)
NEMA MG1-1987 Motors and Generators

Underwriters Laboratories

Underwriters Laboratories has developed an elaborate system of standards for safety that is the basis for their evaluation, listing, and labeling of products. The UL label and listing have been recognized as a symbol of safety by testing and evaluation. They have a high degree of credibility among state, county, and municipal authorities and is world recognized as the Factory Mutual Label.

UL listing and acceptance is documented in product directories. That is when a company has a specific type of product, as indicated by the listing identification listed by UL, it is indicated in a product directory. Two product directories of greatest interest are *Electrical Construction Materials* and the *Hazardous Location Equipment*.

The UL standards program involves industry and other interested groups. The initial development of a standard is open to all interested parties. The review process uses the ANSI canvass method which involves balanced representation of interests and appeal mechanisms and a public periodic review. UL also performs studies and fact-finding reports including a number of such reports on hazardous area equipment.

The UL safety standards of interest include

UL 674 Electric Motors and Generators for Use in Hazardous (Classified) Locations
UL 1277 Electrical Power and Control Tray Cables with Optional Optical Fiber Members
UL 1604 Electrical Equipment for Use in Hazardous (Classified) Locations
UL 913 Intrinsically Safe Apparatus and Associated Apparatus for Use in Class I, II, and III Division 1 Hazardous Locations
UL 96A Installation Requirements for Lightning Protection Systems

Another organization involved in testing for safety and provides third-party certification is the Factory Mutual (FM) Research Corporation. Its *Loss Prevention Data,* reports, and studies provide excellent data for electrical and process control safety. Furthermore, the FM Loss Prevention Manual is a useful safety document. Equipment listing and labeling are also a part of the FM Research Corporation's business.

Listing and certification is also available from Met Electrical Testing Company, Inc., 916 W. Patapsco Ave., Baltimore, Maryland, and ETL Testing Laboratories Inc. Industrial Park, Cortland, New York.

U.S. Department of Labor—Occupational Safety and Health Administration

The Occupational Safety and Health Act of 1970 was responsible for the development of OSHA. OSHA's purpose, coverage, and standards are indicated in the following pages, which are extracted from All About OSHA (1985) (Revised) OSHA 2056. This and other publications are available at U.S. government bookstores.

OSHA develops and publishes standards, conducts hearings and reviews on these standards, reviews appeals and proposed variances to these standards, inspects accident and illness records, performs workplace inspections and accident investigations, issues citations and penalties, responds to appeals by employees or employers, conducts reviews relevant to these appeals, and provides consultations, training, and education.

The OSHA standards, which are published in the *Federal Register,* include Design Safety Standards for Electrical Systems, 29CFR Part 1910 Electrical Standards, covering

1910.302 Electrical Utilization Systems
1910.303 General Requirements
1910.304 Wiring Design and Protection
1910.305 Wiring Methods and Components and Equipment for General Use
1910.306 Specific Purpose Equipment and Installations
1910.307 Hazardous (Classified) Locations
1910.308 Special Systems

This OSHA standard closely parallels the *NEC* and NFPA 70E Electrical Safety Requirements for Employee Workplaces, Part I, Installation Safety Requirements. It is primarily to protect people in the workplace and does not cover all the topics in the *NEC*. Certain sections of the standard are retroactive and apply to facilities installed before 1972. They include identification of equipment, guarding, grounding, overcurrent protection, and hazardous locations.

OSHA also has a standard entitled Electrical Safety—Related Work Practices, 29CFR Part 1910 Aug 6, 1990 covering

1910.332 Training
1910.333 Selection and Use of Work Practices
1910.334 Use of Equipment
1910.335 Safeguards for Personnel Protection

This document closely parallels NFPA 70E Part II Safety Related Work Practices. It covers training, deenergizing equipment, lockout and tagging, bypassing interlocks, working on overhead lines, test equipment, warning signs, protective equipment, and other topics.

Other OSHA standards include

Safety Testing or Certification of Certain Workplace Equipment and Materials, 29CFR Parts 1907, 1910, 1935, and 1936
Control of Hazardous Energy Source (Lockout and Tagout) 29CFR Part-SEPT 1, 1990
Electrical Standards for Construction, 29CFR Part 1926
Process Safety Management of Highly Hazardous Chemicals: Proposed Rulemaking-July 17, 1990

Process Safety Management was prepared in response to recent explosions and fires involving chemical processes and includes sections on many topics relating to safety.

The proposed rules for a process management system covering the following items:

Process safety information
Process hazard analysis
Training
Contractors
Prestartup safety review
Mechanical integrity
Hotwork permits
Management of change
Incident investigation
Emergency planning and response
Compliance safety audit

The background discussion includes a statement that OSHA must depend on national consensus standards and industry standards to support citations. Electrical classification, ventilation, and safety interlocks, topics covered in this text, are part of the required information pertaining to the process.

INTERNATIONAL STANDARDS

The U.S. government has a strong interest in international standards. Furthermore, since the world is moving to an international market, it will be difficult for many industries to survive without having an international perspective.

Global standardization has strong commercial emphasis. Therefore, it is necessary to know international rules to sell and compete in a world market. There are also safety reasons to study international standards: These international standards offer different approaches and practices to process safety that we should consider. They offer new possibilities and opportunities. For instance:

Sale of U.S. products overseas to open new markets
Knowledge of safety practices in U.S.-owned plants outside of the United States
Evaluation of safety of process equipment made overseas and used in the United States
Consideration of overseas safety technology for application to U.S. codes, standards, and practices

The major organization for worldwide electrical standards is the IEC, which was founded in 1906 and includes approximately 40 member countries. The IEC is responsible for electrical standards in many areas; in hazardous locations, the standards are of special interest because of the major differences between the IEC standards and U.S. practice.

Technical committee TC31 is responsible for hazardous locations. It includes these subcommittees:

S/C 31A Flamproof enclosures
S/C 31B Sand-filled apparatus
S/C 31C Increased safety apparatus
S/C 31D Pressurization and associated techniques
S/C 31G Intrinsically safe apparatus
S/C 31H Apparatus for use in the presence of ignitable dust
S/C 31J Classification of hazardous areas and installation requirements
S/C 31K Encapsulation

The standards developed by the IEC for hazardous locations include

79-0 General requirements
79-1 Construction and test of flameproof enclosures of electrical apparatus
79-1 Amendment 1
79-1A Method to test for ascertainment of maximum experimental safe gap
79-2 Electrical apparatus, type of protection p
79-3 Spark test apparatus for intrinsically safe circuits
79-4 Method to test for ignition temperature
79-5 Sand-filled apparatus
79-5A Sand-filled apparatus
79-6 Oil-immersed apparatus
79-7 Construction and test of electrical apparatus, type of protection e
79-8 Classification of maximum surface temperatures
79-9 Marking
79-10 Classification of hazardous areas
79-11 Construction and test of intrinsically safe and associated apparatus
79-12 Classification of mixtures of gases or vapors with air according to their maximum experimental safe gaps and minimum igniting currents

79-13 Construction and use of rooms or buildings protected by pressurization

Not all members of the IEC accept these standards in their entirety. Individual countries have their own standards and practices.

The European Electrotechnical Committee for Standardization (CENELEC) is an organization within the European community aimed at harmonizing the standards and practices of countries within the common market and developing common standards based on the IEC standards. The CENELEC standards include

EN 50014 General requirements
EN 50015 Oil immersion
EN 50016 Pressurized apparatus
EN 50017 Powder filling
EN 50018 Flameproof enclosure
EN 50019 Increased safety
EN 50020 Intrinsic safety

Significant differences in North American and IEC countries exist in basic terminology, codes, requirements, and practices in hazardous locations. For example, in IEC countries classified locations are referred to as zones instead of divisions. The following is a comparison of U.S. and German classification terminology:

United States (NEC)	Germany
Class I Division 1	Zone 0
	Zone 1
Class I Division 2	Zone 2
Class II Division 1	Zone 10
Class II Division 2	Zone 11

Division 1 consists of two zones: Zone 0 a higher risk zone where flammable concentrations can exist continuously or for long periods of time, and Zone 1, a lower risk location where flammable concentrations are likely to occur in normal operation. Zone 0 represents an important distinction between North American and IEC practice. Europe has two levels of intrinsic safety: type ia, two fault, which is required in Zone 0, and type ib, single fault, which is acceptable in Zone 1.

Differences in terminology are also applied to gas and vapor groups. For instance, the IEC gas group IIA is similar to U.S. group D, whereas IIB is similar to group C, and IEC group IIC is similar to Groups A and B.

The marking labels on European equipment follows IEC standards. The first symbol is EX, indicating the equipment is intended for a flammable gas or vapor location. The second symbol indicates the types of protection, which can be any of the following:

Flameproof (similar to explosion proof)	d
Pressurized enclosure	p
Intrinsic safety	ia or lb
Oil immersion	o
Increased Safety	e
Sand filled	q
Special protection	s

The last letter indicates the safe surface temperature similar to *NEC* table 300-3 (b) except fewer temperature classes. An example of a marking is

Ex d IIB T1

This indicates a flameproof enclosure, acceptable for Group IIB gases or vapors, and safe for a gas or vapor whose autoignition temperature is greater than 450°C (T1).

CONCLUSION

This chapter has described the many national and international standards that provide excellent resources in the design, installation, and maintenance of electrical and process control facilities in chemical processes. These standards provide guidance to engineers and other safety personnel as to the right thing to do. They also provide legal credibility for the actions. Complying with these standards puts one on good engineering and legal footing. However, all standards must be applied with good engineering judgment.

Standards are also dynamic and change as engineering technology changes. It is important to keep up with changes and to participate in standards programs by providing input. Each *NEC* cycle, there are thousands of proposals aimed at updating and improving the standard, and this is as it should be.

10

Safety in Maintenance

Maintenance is the process of repairing, correcting, restoring, and modifying equipment for the purpose of maintaining the operability, reliability, and safety of equipment and systems. In chemical-process facilities, maintenance of electrical and process control systems is an important safety issue. A significant percentage of safety incidents are attributed to improper or no maintenance of these systems. In one incident, a plant stopped cleaning substations, and after a time, dust accumulated in switch gear enclosures. During a severe rainstorm, the dust accumulation became moist and caused a voltage flashover from the switch gear bus bar across the insulation and to the switch gear enclosure. The flashover caused the circuit breakers to operate, which eventually shut down the entire plant.

Maintenance is especially important in chemical processes because of the harsh environment, the complexity and size of electrical and process control systems, the requirement to keep the processes operating continuously, and the hazardous nature (fire, explosion, toxicity, and pollution) of the process materials.

In general, maintenance is required because of

Equipment wear
Deterioration of enclosures
Changes—redesign and recalibration
Corrections to original installation

Maintenance is also required to correct operational problems, for instance, when a loop is out of control, a high temperature alarm occurs for no reason, or a motor shuts down.

Examples of maintenance work are

Replacing motor bearings
Repairing a corroded junction box

Replacing a damaged limit switch on a control valve
Replacing damaged electrical cable on a flow transmitter

Plants and organizations have different philosophies on maintenance practice. Some facilities provide maintenance only when there is a shut down, failure, accident, fire, spill, or release to the air. That is, these operations only respond to problems instead of preventing them. This type of maintenance is difficult to understand in today's safety and environmentally conscious chemical industry. The system is in contrast to an engineered and managed program aimed at prevention.

PREVENTIVE MAINTENANCE

A well-engineered and managed preventive maintenance program includes the following:

- Surveys and inspections
- Scheduled maintenance
- Recordkeeping of problems and other critical data
- Spare parts program
- Training of personnel, especially in the use of new equipment and systems, maintenance and safety practices, codes and standards
- Documentation program to update one-line diagrams, schematics, loop diagrams, PLC and process computer software, inspection checklist, Electrical Area Classification Drawings, and other critical data
- Use of outside testing organizations to perform studies and test selected equipment and systems, for instance, transformer tests, setting substation breakers, analyzer calibration and service
- Participation in the design of new or modified process facilities to provide maintenance input

DESIGN FOR MAINTAINABILITY

Designing new facilities for maintainability can do much to reduce or eliminate problems. In particular, the design of new or modified facilities should consider maintainability and reliability in the following ways.

Avoid Hotwork. Design systems capable of servicing without working on or near energized equipment or wiring (hotwork). This requires providing isolation switches, back-up systems, manual controls, isolation and separation of systems, isolation valves, control circuit switches, and so on, wherever practical and reasonable, but especially where personnel or process safety is at risk. Hotwork is not only risky for the people working on the energized equipment but also to the process. A mistake could shut down the process and possibly damage critical electrical or electronic equipment.

Enclosure Design. Specify enclosures suitable for the environment (NEMA 4, NEMA 12, etc.) Water, dust humidity, and chemical corrosion are major factors in electrical system failures. In some extremely corrosive conditions, plastic enclosure should be considered. Electrical heaters to remove condensation can be provided in switch gear, motor, and other enclosures. Where explosionproof equipment is required, avoid the ground surface cover design unless an O ring seal can be provided. Purging and pressurization can also minimize exposure to outside air.

Keep It Simple. Design simple systems, especially interlock systems. Many control systems are overdesigned and overinterlocked. It is especially easy to overdesign and overcontrol in an attempt to protect against all possible failures, especially with the flexibility of PLC and process computer software interlock capability. Overdesigned systems, however, lose their credibility. In addition, the important interlocks are lost in the crowd. In one situation, a motor control circuit was so overinterlocked it could not be started.

Use Proven Equipment. Avoid using new, complex, and highly specialized equipment unless there is no alternative. Use instruments, controls, and electrical equipment that is familiar to the plant maintenance department and that has a record of good reliability and performance. New types of equipment result in needing additional training, spare parts, and service manuals. This should not deter trying new designs to provide improved performance over existing equipment.

Provide Adequate Working Room. There should be adequate working room around electrical equipment, in front of panels, and within electrical enclosures. Sections 110-16, 17, and 34 of the *NEC* describe the minimum wiring distances and the rules for guarding equipment. Congested wiring in control room panels is a safety concern in many situations. Therefore, inside equipment there should be ample room for external cables and for removing components for testing and replacement.

ELECTRICAL AND PROCESS CONTROL SYSTEM MAINTENANCE PRACTICES

Before working on any equipment or system, the work must be coordinated with the operations department. Specifically, the personnel assigned to the task must have a clear understanding of their work, must be trained and knowledgeable about the equipment and system, must have up-to-date drawings and manufacturers' data on the equipment, and must have correct and inspected tools and test equipment. Finally, the location where the maintenance is to be done must be free of hazards. If the maintenance area is on or near energized circuits or equipment, special hotwork procedures are necessary. If the area is classified as hazardous, special precautions are necessary.

SURVEYS AND INSPECTIONS

Electrical Enclosures Integrity

Electrical enclosures protect electrical equipment, components, and wiring from the outside environment, and their integrity is essential to reliable system operation. An internal and external visual inspection will indicate corrosion or leakage into the enclosure. Inspection of gasketing and covers will ensure they are in place.

Cable and Connection Integrity

Connections, especially high-current power connections, are the weakest link in the wiring system and can cause hot spots and fires. Therefore, cables and cords need to be inspected for damage during initial installation or when in use. Aluminum conductor connections and mechanical connections especially should be checked. Infrared thermometers or thermal imaging systems are excellent tools to find hot spots resulting from poor contact. Thermometers are handheld, portable devices that are aimed at wiring connections a *short distance* away and provide a readout of temperatures and hot spots. Thermal imaging systems produce an x-ray type of photograph that identifies hot spots over a large area of equipment or wiring. Infrared thermometers can be used routinely by maintenance personnel; thermal imaging photos are usually done periodically by outside specialists.

Grounding Inspections

Grounding inspections should begin at the substation and proceed through the entire electrical system. All metallic electrical enclosures, conduit, cable tray, and so on, should be bonded together and back to the power transformer neutral. The transformer neutral should be system grounded either solidly or through a resistor to building steel or earth. All grounding and bonding connections must be reliable and not damaged or corroded.

Circuit and Equipment Protection

Verify that the rating of circuit breakers and fuses are per the plant one-line diagrams and substation breakers are on the settings as indicated in the breaker specifications and coordination curves.

Electrical and Electronic Equipment Cooling

Verify that fans on electronic equipment are operating and that all filters are clean. Loss of cooling air to this type of equipment can cause shut down of the electronics and major component failures. Verify that any cooling to force ventilated motors is working and that the loss-of-air interlocks are working.

Polychlorinated Biphenyl (PCB) Inspections

Failure to follow EPA regulations concerning polychlorinated bipheyls (PCB) in substation transformers and power capacitors can result in serious fines. Verify that any PCB-contaminated equipment complies with the EPA Electrical Rules and Fires Rules and that the proper monitoring and record-keeping procedures are in place. Transformer fires involving PCBs can be catastrophic. If PCB transformers are present, verify that they are protected according to IEEE standards.

Process Control and Electrical Rooms

Inspect process control and electrical rooms or buildings to ensure that there are no openings to permit outside air to enter.

TESTING

Electrical Power Systems

Testing electrical transformers, cables, motors, protective relaying, and switch gear has been established by IEEE standards. Test frequency and methodology are discussed in these standards as well as in NFPA 70B and the manufacturers' instructions. Test meters, test methodology, safety procedures, and documentation are essential elements of a test program. Outside testing companies are usually used to perform transformer oil testing, protective relay testing, breaker service and testing, substation earth resistance testing, and other tests. These companies have the specialized test equipment and expertise to perform the tests, but it is essential to monitor their work and the test results closely.

Acceptance Testing of Equipment

Testing electrical equipment at the manufacturer's shop before shipping the equipment is an excellent opportunity to verify the acceptability of the equipment and establish its performance. For instance, full load tests on a large motor can establish its efficiency, power factor, and other important characteristics that can be used when the motor is running. Insulation test results can serve as a benchmark for evaluating the condition of the motor insulation years later. The maintenance department should have input as to the acceptance tests that will be performed on the equipment.

Testing Instrumentation and Control Systems

The in-place testing of instrument primary measuring elements is limited because the output signal is dependent on the actual process fluid. For instance, to check the output signal of load cells, it may be necessary to add calibrated dead weights on the vessel, but the vessel must be out of service. It may be possible to check thermocouples in place with a known temperature medium. Testing process-connected instrumentation is also complicated by possible contact with process fluids, which can be toxic, corrosive, and flammable, and by the process location, which has the same exposure. Instrumentation process connections must be designed for reliability and maintainability with all flanges, isolation valves, and connections matched to process piping in terms of pressure, temperature, and corrosion rating and the type of connection. Isolation valves and flushing connections need to be provided to minimize the possibility of exposure to process fluids. Process connections should be inspected periodically for leaks, damage, or corrosion. Plant shut downs provide the opportunity to simulate signals and remove instrumentation for testing and calibration.

Interlock system testing should also be accomplished during shut downs. On-line testing of interlocks requires carefully controlled and managed application of manual controls and bypasses and coordination with operations. Up-to-date schematics, loop diagrams, PLC, and process computer software for the interlocks are essential.

Electrical Measurements

Tests on electrical equipment provide useful information on the condition of the electrical equipment and process conditions. It is important to know load conditions. For instance, motor-running current should be compared to the motor nameplate data. System voltages should be measured periodically to ensure that the transformers are connected properly and are on the correct tap. An incident occurred recently in which a 120-volt secondary winding on a control power transformer was incorrectly connected for 240 volts. The transformer was incorrectly connected for 240 volts. The transformer served the field winding of a large synchronous motor that burned it open, resulting in a significant loss. Electrical power to process heaters should be measured and recorded to ensure that the heater elements are properly connected. Control power measurements are also essential to prevent shut down of these critical systems due to overloads. This includes dc power measurements.

REFERENCES

Crouse Hinds Co. Suggestions For Installation and Maintenance of Electrical Equipment For Use In Hazardous Areas. Bulletin 2911.

Dudor, J.S. 1989. Application and Use of Inspection Checklists for Factory and Field Inspection of Electrical Equipment. *IEEE Transactions on Industry Applications* 25 (5, Sept./Oct.)

National Fire Protection Association. 1983. *NFPA 70B Recommended Practice for Electrical Equipment Maintenance.*

11

Safety in Work Practices

ELECTRICAL AND PROCESS CONTROL SAFETY WORK PRACTICES

The topic of safe work practices is closely related to that of safe maintenance. NFPA 70E indicates that maintenance involves restoring or preserving the condition of electrical facilities; whereas work practices refers to the procedures for personnel who work near, on, or with equipment and systems. They could be viewed as what to do (maintenance) and how to do it (work practices).

Accident reports indicate that a significant percentage of incidents are attributable to unsafe acts or work practices. Unsafe work practices can be the result of insufficient training, improper lockout and tagging, inadequate clearance of overhead power lines, inadequate illumination, inadequate working space, damaged cords and plugs, inadequate protective equipment, and so on. There are numerous examples.

An operator tried to open a disconnect switch for a large extruder drive motor. The switch, an isolation disconnect, was not rated to open underload and blew apart when it was opened underload. In another case, a multitester was used to check voltage to a motor to ensure it was locked out. With the meter leads connected to 460 volts, the meter scale was inadvertently turned to the resistance scale, and the meter short circuited the 460 volts. (There are meters available that are protected against this circumstance.) The meter lead wires, fortunately, acted as a fuse and flashed open the circuit before significant damage occurred. These and other such incidents clearly indicate that work practices are a safety issue.

In chemical-process facilities lockout of electrical facilities, isolation valves, and process equipment are safety concerns. A report of the recent explosion at the Phillips Petroleum facility near Houston that killed 23 people indicated that a bottom valve was probably left open when a unit started, even though there were three separate lockout mechanisms for this valve.

OSHA has recognized the importance of safe work practices by developing the standard *Electrical Safety: Related Work Practices.* The standard is based on NFPA 70E *Part II, Safety Related Work Practices.* It is performance oriented and covers work performed near or on exposed energized parts of electrical equipment. In particular it covers

Training requirements
Lockout and tagging
Overhead power lines
Adequate illumination
Working space
Plug and connected equipment
Overcurrent protection
Protective equipment

The standard is directed at unqualified as well as qualified people. Qualified people are those who are well acquainted with and thoroughly familiar with the equipment.

Training requirements include safety-related work practices as described in the standard and other safety practices particular to the work place. Classroom and job training are acceptable.

The elements of lockout and tagging are discussed and further developed in the OSHA standard, *Control of Hazardous Energy Sources (Lockout and Tagout).* The purpose of lockout and tagout is to prevent equipment from being inadvertently energized when someone is working in or on equipment. If equipment is locked out, it is dead as far as external energy sources are concerned. This refers to the energization of electrical and mechanical devices. Lockout is the placement of a device that isolates the energy source. Most plants have had lockout procedures, but now the OSHA standard requires a written lockout and tagout program and that the procedure be developed, documented, and practiced. Work on cord and plug connected equipment is not included.

Tagout is the placement of a tag on an energy-isolating device, like a disconnect, according to an established procedure, indicating that the device, which usually has a warning marked on it, cannot be operated until the tag is removed. Typical warnings include "Do Not Start," "Do Not Open," "Do Not Close," or "Do Not Operate." Lockout is always preferred, but when it is not possible, tagout is permitted if it can be demonstrated to provide employee protection. If the equipment is capable of lockout, it should be locked out.

Periodic inspections should be instituted to ensure that the procedures are being followed. Employee training and communication of the procedures and techniques are required. Verification of isolation before working on the equipment is required. Some plants provide a test position on a start push button at the motor to verify the motor will not start. The motor power disconnect is locked open at the MCC, and the test push button is operated to ensure the correct disconnect was opened. The test push-button operation bypasses all interlocks to ensure that an interlock did not prevent the motor instead of the disconnect from starting. Motor push buttons are not acceptable lockout devices; opening a disconnect in the power circuit to the motor is required. The procedure for removal of lockout and tagout devices must be enforced.

A lockout and tagout program is essential for all types of equipment, electrical and instrument, even low-voltage power distribution breakers. A case history illustrates the consequences of failing to follow tagout practices on electrical panels. A UL-listed light was to be installed in a paint booth to replace a standard fixture. The electrician turned off the breaker that energized the circuit for the fixture and proceeded to replace the standard fixture, but the wiring to the fixture was "hot" and caused sparking. That sparking ignited a solvent in the spray booth and caused a fire that burned down the entire building.

Other items in the OSHA standard include

- Only qualified people may defeat electrical interlocks.
- Overhead lines are to be deenergized or grounded and other protective measures taken.
- All test equipment should be visually inspected.
- Test equipment is to be appropriate for the job.
- Appropriate protective equipment is to be used.
- Warning signs to alert employees of electrical hazards are to be provided.
- Insulated tools are to be used near energized parts.
- Fuse-handling tools are to be used.
- Overcurrent protection may not be modified.
- Load-rated switches are required if they are intended to open underload.

Operators should not stand directly in front of power switch enclosures during operation. The operator should stand on the handle side of the enclosure and operate the handle with the hand closest to it. This will keep the operator out of the direct path of any flashes. The operator's face should always be turned away from the door when the switch is operated. The operator should also wear safety glasses, gloves, and protective clothing. If test leads enter the enclosure, the door should be closed as much as possible. Before closing a switch or breaker, verify that the load is not inadvertently short circuited.

HOT WORK

Hot work means working on or in close proximity to exposed energized electrical equipment or wiring. The term is usually applied to working in energized substations, high-voltage switch gear, or MCCs, but it should also be applied to working on energized instrumentation, interlocks, and control systems.

The type of safety risks is different for instrumentation and power. In substations, the concern is high voltage and high currents, and personnel safety is a major issue. Shock, electrocution, and severe burns are possible. High-current arcs can radiate deadly temperatures 4–6 feet away. There is also the risk of explosions, fires, shut down, and major damage to electrical power facilities. With instrument and control system hot work, the primary concern is process safety. Personnel safety, however, may be a concern at voltages over 30 volts and in hazardous locations. Hot work should always be avoided by proper design of isolation switches or redundant systems or by waiting for a shut down of facilities. There may, however, be some situations where hot work is necessary, so plants should have established hot work policies and procedures.

Hot work should be classified according to the risks involved and the protective measures required. The risk involved in a 13.8 kilovolt, 10-MVA substation is considerably higher than in a 480-volt MCC. Hot work on interlocks in a large, continuous-process petrochemical facility is a higher process risk than in a material-handling system processing plastic pellets. The protective measures include the required expertise of the personnel performing the hot work, the number of people involved (two as a minimum), the protective clothing and tools, standby emergency groups, and the level of management approval of hot work. Coordination with the operating department is always a requirement.

MAINTENANCE IN HAZARDOUS (CLASSIFIED) LOCATIONS

Periodic inspection of classified locations is essential for chemical-process safety. The facilities required in these locations must be inspected and maintained to comply with Article 500 of the *NEC*. The first step in inspecting a classified location is to review the Electrical Area Classification drawings to ensure they are up to date. This requires knowing the flammable materials (e.g., acetone and propane). Verify that the class, group, and T numbers match the materials processed. Check that all explosionproof or dust-ignition-proof enclosures are marked with the correct class, group, and T number. T-number marking is only required on heat-generating equipment (e.g., lights, motors, heaters). Verify that all explosion seals are installed and that enclosures are not damaged, corroded, or opened. Purged and pressurized enclosures should have flow or pressure indication and loss of pressure alarms or interlocks. Bolted covers on explosionproof enclosures should have all the bolts installed and tight. All threaded joints and fittings should be tight. Intrinsically safe systems should have the cabling and wiring systems separated as indicated in the manufacturer's control drawing. Power or other control wiring should be in the same cable tray or raceway with IS wiring. All metal enclosures, conduct, cable tray, and so on, should be bonded together, and the transformer neutral should be system grounded to building steel or earth. External or internal bonding jumpers are required around sealtite. Locknut and double locknut types of bonding connections are not permitted.

HOT WORK IN HAZARDOUS (CLASSIFIED) LOCATIONS

Hot work in any location is to be avoided, but especially so in classified locations where an arc or spark has the potential for fire and explosion. This topic is not addressed in national or international standards.

The following is one approach to hot work in classified locations for various types of systems. Manufacturers of explosionproof equipment provide warnings on the enclosures indicating that the cover is not to be removed if the location is hazardous unless the wiring is deenergized. In some continuous process operations, however, hot work in classified locations may be necessary.

The risks associated with low-voltage instrumentation are considerably less than with power equipment. Intrinsically safe systems have distinct advantages over other systems because IS barriers limit the arc energy in the classified location. Nonincendive systems also have an advantage in Division 2 locations while Purged and pressurized enclosures provide some degree of protection for a period of time, depending on the flow capacity of the purge system.

Each plant must choose its own practices, but the following are options based on the use of combustible gas detectors. (Combustible gas detectors are used to provide protection for nonlisted equipment in Canada.)

Hot Work Options in Class I Locations

Division 2 Use combustible gas detectors set and tested to alarm at 25%. Stop hot work on alarm.

Division 1 Use combustible gas detectors as above with purged and pressurized enclosures. IS systems can be worked hot. All other equipment must be deenergized.

REFERENCES

McClung, B., and J.M. Gallagher. 1987. Electrical Hot Work. *Electrical Construction and Maintenance Magazine* (Feb.)

National Fire Protection Association. 1988. *NFPA 70E Standard for Electrical Safety Requirements for Employee Workplaces.*

OSHA. 1989. Control of Hazardous Energy Sources (Lockout and Tagout); Final Rule. *Federal Register* 29CFR Part 1910 (Sept. 1).

OSHA. August 6, 1990. Electrical Safety: Related Work Practices; Final Rule. *Federal Register* 29CFR Part 1910. 331–335.

Appendix

The following information is extracted from the OSHA publication, *An Illustrated Guide to Electrical Safety*, U.S. Department of Labor, OSHA–1983 (Revised), OSHA 3073.

It provides additional information complementary to chapters 3, 4, and 5.

AN ILLUSTRATED GUIDE TO ELECTRICAL SAFETY

U.S. Department of Labor, Occupational Safety and Health Administration, 1983 (Revised), OSHA 3073.

§1910.307 Hazardous (classified) locations.

(a) Scope. This section covers the requirements for electric equipment and wiring in locations which are classified depending on the properties of the flammable vapors, liquids or gases, or combustible dusts or fibers which may be present therein and the likelihood that a flammable or combustible concentration or quantity is present. Hazardous (classified) locations may be found in occupancies such as, but not limited to, the following: aircraft hangers, gasoline dispensing and service stations, bulk storage plants for gasoline or other volatile flammable liquids, paint-finishing process plants, health care facilities, agricultural or other facilities where excessive combustible dusts may be present, marinas, boat yards, and petroleum and chemical processing plants. Each room, section or area shall be considered individually in determining its classification. These hazardous (classified) locations are assigned six designations as follows:

Class I, Division 1
Class I, Division 2
Class II, Division 1
Class II, Division 2
Class III, Division 1
Class III, Division 2

For definitions of these locations see §1910.399(a). All applicable requirements in this subpart shall apply to hazardous (classified) locations, unless modified by provisions of this section.

- THE FOLLOWING DISCUSSION PROVIDES A GENERAL OVERVIEW OF THE GUIDELINES CONTAINED IN THE NATIONAL ELECTRICAL CODE, CHAPTER 5. ALSO, HIGHLIGHTS AND SUMMARY INFORMATION ARE PRESENTED TO AID IN UNDERSTANDING DESIGN CONCEPTS AND EQUIPMENT SELECTION. SEVERAL REFERENCES ARE MADE TO NEC ARTICLES 500, 501, 502, AND 503. CAREFUL STUDY OF THESE AND THEIR ASSOCIATED ARTICLES SHOULD PRECEDE ANY DESIGN DEVELOPMENT ACTIVITIES.

HAZARDOUS LOCATIONS

HAZARDOUS LOCATIONS ARE AREAS WHERE FLAMMABLE LIQUIDS, GASES, OR VAPORS, OR COMBUSTIBLE DUSTS EXIST IN SUFFICIENT QUANTITIES TO PRODUCE AN EXPLOSION OR FIRE. IN HAZARDOUS LOCATIONS, SPECIALLY DESIGNED EQUIPMENT AND SPECIAL INSTALLATION TECHNIQUES MUST BE USED TO PROTECT AGAINST THE EXPLOSIVE AND FLAMMABLE POTENTIAL OF THESE SUBSTANCES.

HAZARDOUS LOCATIONS ARE CLASSIFIED AS CLASS I, CLASS II, OR CLASS III, DEPENDING ON WHAT TYPE OF HAZARDOUS SUBSTANCE IS OR MAY BE PRESENT. IN GENERAL, CLASS I LOCATIONS ARE THOSE IN WHICH FLAMMABLE VAPORS AND GASES MAY BE PRESENT. CLASS II LOCATIONS ARE THOSE IN WHICH COMBUSTIBLE DUSTS MAY BE FOUND. CLASS III LOCATIONS ARE THOSE IN WHICH THERE ARE IGNITIBLE FIBERS AND FLYINGS.

EACH OF THESE CLASSES IS DIVIDED INTO TWO HAZARD CATEGORIES, DIVISION 1 AND DIVISION 2, DEPENDING ON THE LIKELIHOOD OF THE PRESENCE OF A FLAMMABLE OR IGNITIBLE CONCENTRATION OF A SUBSTANCE. DIVISION 1 LOCATIONS ARE DESIGNATED AS SUCH BECAUSE A FLAMMABLE GAS, VAPOR, OR DUST IS NORMALLY PRESENT IN HAZARDOUS QUANTITIES. IN DIVISION 2 LOCATIONS, THE EXISTENCE OF HAZARDOUS QUANTITIES OF THESE MATERIALS IS NOT NORMAL, BUT THEY MAY OCCASIONALLY EXIST EITHER ACCIDENTALLY OR WHEN MATERIAL IN STORAGE IS HANDLED. IN GENERAL, THE INSTALLATION REQUIREMENTS FOR DIVISION 1 LOCATIONS ARE MORE STRINGENT THAN FOR DIVISION 2 LOCATIONS.

ADDITIONALLY, CLASS I AND CLASS II LOCATIONS ARE ALSO SUBDIVIDED INTO GROUPS OF GASES, VAPORS, AND DUSTS HAVING SIMILAR PROPERTIES.

TABLE 3 SUMMARIZES THE VARIOUS HAZARDOUS (CLASSIFIED) LOCATIONS. THE DEFINITIONS OF THE LOCATIONS ARE GIVEN IN PARAGRAPH 1910.399(a) OF THE STANDARD.

TABLE 3. SUMMARY OF CLASS I, II, III HAZARDOUS LOCATIONS

CLASSES	GROUPS	DIVISIONS 1	DIVISIONS 2
I GASES, VAPORS, AND LIQUIDS (ART. 501)	A: ACETYLENE B: HYDROGEN, ETC. C: ETHER, ETC. D: HYDROCARBONS, FUELS, SOLVENTS, ETC.	NORMALLY EXPLOSIVE AND HAZARDOUS	NOT NORMALLY PRESENT IN AN EXPLOSIVE CONCENTRATION (BUT MAY ACCIDENTALLY EXIST)
II DUSTS (ART. 502)	E: METAL DUSTS (CONDUCTIVE* AND EXPLOSIVE) F: CARBON DUSTS (SOME ARE CONDUCTIVE,* AND ALL ARE EXPLOSIVE) G: FLOUR, STARCH, GRAIN, COMBUSTIBLE PLASTIC OR CHEMICAL DUST (EXPLOSIVE)	IGNITABLE QUANTITIES OF DUST NORMALLY IS OR MAY BE IN SUSPENSION, OR CONDUCTIVE DUST MAY BE PRESENT	DUST NOT NORMALLY SUSPENDED IN AN IGNITIBLE CONCENTRATION (BUT MAY ACCIDENTALLY EXIST). DUST LAYERS ARE PRESENT.
III FIBERS AND FLYINGS (ART. 503)	TEXTILES, WOODWORKING ETC. (EASILY IGNITABLE, BUT NOT LIKELY TO BE EXPLOSIVE)	HANDLED OR USED IN MANUFACTURING	STORED OR HANDLED IN STORAGE (EXCLUSIVE OF MANUFACTURING)

*NOTE: ELECTRICALLY CONDUCTIVE DUSTS ARE DUSTS WITH A RESISTIVITY LESS THAN 10^5 OHM-CENTIMETER.

(b) Electrical installations. Equipment, wiring methods, and installations of equipment in hazardous (classified) locations shall be intrinsically safe, or approved for the hazardous (classified) location, or safe for the hazardous (classified) location. Requirements for each of these options are as follows:

(1) Intrinsically safe. Equipment and associated wiring approved as intrinsically safe shall be permitted in any hazardous (classified) location for which it is approved.

(2) Approved for the hazardous (classified) location.

(i) Equipment shall be approved not only for the class of location but also for the ignitible or combustible properties of the specific gas, vapor, dust, or fiber that will be present.

NOTE: NFPA 70, the National Electrical Code, lists or defines hazardous gases, vapors, and dusts by "Groups" characterized by their ignitible or combustible properties.

(ii) Equipment shall be marked to show the class, group, and operating temperature or temperature range, based on operation in a 40 degrees C ambient, for which it is approved. The temperature marking may not exceed the ignition temperature of the specific gas or vapor to be encountered. However, the following provisions modify this marking requirement for specific equipment:

(a) Equipment of the non-heat-producing type, such as junction boxes, conduit, and fittings, and equipment of the heat-producing type having a maximum temperature not more than 100 degrees C (212 degrees F) need not have a marked operating temperature or temperature range.

(b) Fixed lighting fixtures marked for use in Class I, Division 2 locations only, need not be marked to indicate the group.

(c) Fixed general-purpose equipment in Class I locations, other than lighting fixtures, which is acceptable for use in Class I, Division 2 locations need not be marked with the class, group, division, or operating temperature.

(d) Fixed dust-tight equipment, other than lighting fixtures, which is acceptable for use in Class II, Division 2 and Class III locations need not be marked with the class, group, division, or operating temperature.

(3) Safe for the hazardous (classified) location. Equipment which is safe for the location shall be of a type and design which the employer demonstrates will provide protection from the hazards arising from the combustibility and flammability of vapors, liquids, gases, dusts, or fibers.

NOTE: The National Electrical Code, NFPA 70, contains guidelines for determining the type and design of equipment and installations which will meet this requirement. The guidelines of this document address electric wiring, equipment, and systems installed in hazardous (classified) locations and contain specific provisions for the following: wiring methods, wiring connections; conductor insulation, flexible cords, sealing and drainage, transformers, capacitors, switches, circuit breakers, fuses, motor controllers, receptacles, attachment plugs, meters, relays, instruments, resistors, generators, motors, lighting fixtures, storage battery charging equipment, electric cranes, electric hoists and similar equipment, utilization equipment, signaling systems, alarm systems, remote control systems, local loud speaker and communication systems, ventilation piping, live parts, lightning surge protection, and grounding. Compliance with these guidelines will constitute one means, but not the only means, of compliance with this paragraph.

- EQUIPMENT DESIGN

GENERAL-PURPOSE ELECTRICAL EQUIPMENT CAN CAUSE EXPLOSIONS AND FIRES IN AREAS WHERE FLAMMABLE VAPORS, LIQUIDS, AND GASES, AND COMBUSTIBLE DUSTS OR FIBERS ARE PRESENT. THESE AREAS REQUIRE SPECIAL ELECTRICAL EQUIPMENT WHICH IS DESIGNED FOR THE SPECIFIC HAZARD INVOLVED. THIS INCLUDES EXPLOSION-PROOF EQUIPMENT FOR FLAMMABLE VAPOR, LIQUID AND GAS HAZARDS AND DUST-IGNITION-PROOF EQUIPMENT FOR COMBUSTIBLE DUST. OTHER EQUIPMENT USED INCLUDE: NONSPARKING EQUIPMENT, INTRINSICALLY SAFE EQUIPMENT AND PURGED AND PRESSURIZED EQUIPMENT. IN SOME CASES, GENERAL PURPOSE OR DUST-TIGHT EQUIPMENT IS PERMITTED IN DIVISION 2 AREAS.

MANY PIECES OF ELECTRICAL EQUIPMENT INCLUDE CERTAIN PARTS THAT ARC, SPARK, OR PRODUCE HEAT UNDER NORMAL OPERATING CONDITIONS. FOR EXAMPLE, CIRCUIT CONTROLS, SWITCHES, AND CONTACTS MAY ARC OR SPARK WHEN OPERATED. MOTORS AND LIGHTING FIXTURES ARE EXAMPLES OF EQUIPMENT THAT MAY HEAT UP. THESE ENERGY SOURCES CAN PRODUCE TEMPERATURES HIGH ENOUGH TO CAUSE IGNITION. SEE FIGURE 77. ELECTRICAL EQUIPMENT SHOULD NOT BE INSTALLED IN KNOWN OR POTENTIALLY HAZARDOUS LOCATIONS UNLESS ABSOLUTELY NECESSARY. HOWEVER, WHEN ELECTRICAL EQUIPMENT MUST BE INSTALLED IN THESE AREAS, THE SPARKING, ARCING, AND HEATING NATURE OF THE EQUIPMENT MUST BE CONTROLLED.

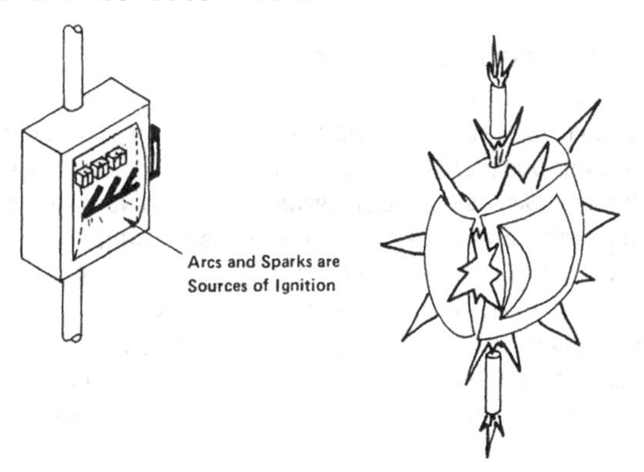
Arcs and Sparks are
Sources of Ignition

IF GENERAL-PURPOSE EQUIPMENT IS USED IN HAZARDOUS LOCATIONS, A SERIOUS FIRE AND EXPLOSION HAZARD EXISTS.

FIGURE 77. EXPLOSION OCCURRING IN GENERAL PURPOSE EQUIPMENT

INSTALLATIONS IN HAZARDOUS LOCATIONS MUST BE: (1) INTRINSICALLY SAFE, (2) APPROVED FOR THE HAZARDOUS LOCATION, OR (3) OF A TYPE AND DESIGN WHICH PROVIDES PROTECTION FROM THE HAZARDS ARISING FROM THE COMBUSTIBILITY AND FLAMMABILITY OF THE VAPORS, LIQUIDS, GASES, DUSTS, OR FIBERS THAT WILL BE PRESENT. INSTALLATIONS CAN BE ONE OR ANY COMBINATION OF THESE OPTIONS. EACH OPTION IS DESCRIBED IN THE FOLLOWING DISCUSSION.

*INTRINSICALLY SAFE

EQUIPMENT AND WIRING APPROVED AS INTRINSICALLY SAFE IS ACCEPTABI E IN ANY HAZARDOUS (CLASSIFIED) LOCATION FOR WHICH IT IS DESIGNED. INTRINSICALLY SAFE EQUIPMENT IS NOT CAPABLE OF RELEASING SUFFICIENT ELECTRICAL OR THERMAL ENERGY UNDER NORMAL OR ABNORMAL CONDITIONS TO CAUSE IGNITION OF A SPECIFIC FLAMMABLE OR COMBUSTIBLE ATMOSPHERIC MIXTURE IN ITS MOST EASILY IGNITIBLE CONCENTRATION.

TO AVOID CONTAMINATING NONHAZARDOUS LOCATIONS, THE PASSAGE OF FLAMMABLE GASES AND VAPORS THROUGH THE EQUIPMENT MUST BE PREVENTED. ADDITIONALLY, ALL INTERCONNECTIONS BETWEEN CIRCUITS MUST BE EVALUATED TO BE SURE THAT AN UNEXPECTED SOURCE OF IGNITION IS NOT INTRODUCED THROUGH OTHER NONINTRINSICALLY SAFE EQUIPMENT. SEPARATION OF INTRINSICALLY SAFE AND NONINTRINSICALLY SAFE WIRING MAY BE NECESSARY TO ENSURE THAT THE CIRCUITS IN HAZARDOUS (CLASSIFIED) LOCATIONS REMAIN SAFE.

*APPROVED FOR THE HAZARDOUS (CLASSIFIED) LOCATION

UNDER THIS OPTION, EQUIPMENT MUST BE APPROVED FOR THE CLASS, DIVISION, AND GROUP OF LOCATION. THERE ARE TWO TYPES OF EQUIPMENT SPECIFICALLY DESIGNED FOR HAZARDOUS (CLASSIFIED) LOCATIONS – EXPLOSION PROOF AND DUST-IGNITION PROOF. EXPLOSION-PROOF APPARATUS IS INTENDED FOR CLASS I LOCATIONS, WHILE DUST-IGNITION-PROOF EQUIPMENT IS PRIMARILY INTENDED FOR CLASS II AND III LOCATIONS. EQUIPMENT APPROVED SPECIFICALLY FOR HAZARDOUS LOCATIONS CARRIES AN UNDERWRITERS' LABORATORIES, INC. (UL), LABEL AND INDICATES IN WHAT CLASS, DIVISION, AND GROUP OF LOCATION IT MAY BE INSTALLED. SEE FIGURE 78. EQUIPMENT APPROVED FOR USE IN A DIVISION 1 LOCATION MAY BE INSTALLED IN A DIVISION 2 LOCATION OF THE SAME CLASS AND GROUP.

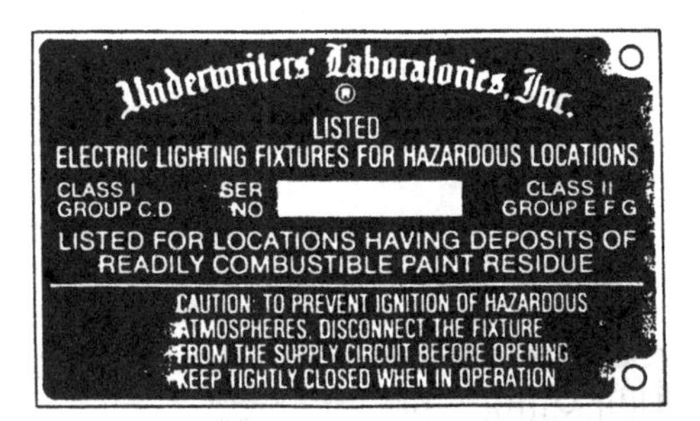

FIGURE 78. LABEL SHOWING APPROVAL FOR USE IN HAZARDOUS (CLASSIFIED) LOCATIONS

EXPLOSION-PROOF EQUIPMENT

GENERALLY, EQUIPMENT INSTALLED IN CLASS I LOCATIONS MUST BE APPROVED AS EXPLOSION-PROOF. SINCE IT IS IMPRACTICAL TO KEEP FLAMMABLE GASES OUTSIDE OF ENCLOSURES, ARCING EQUIPMENT MUST BE INSTALLED IN ENCLOSURES THAT ARE DESIGNED TO WITHSTAND AN EXPLOSION. THIS MINIMIZES THE RISK OF HAVING AN EXTERNAL EXPLOSION OCCUR WHEN A FLAMMABLE GAS ENTERS THE ENCLOSURE AND IS IGNITED BY THE ARCS. SEE FIGURE 79. NOT ONLY MUST THE EQUIPMENT BE STRONG ENOUGH TO WITHSTAND AN INTERNAL EXPLOSION, BUT THE ENCLOSURES MUST BE DESIGNED TO VENT THE RESULTING EXPLOSIVE GASES. THIS VENTING MUST ENSURE THAT THE GASES ARE COOLED TO A TEMPERATURE BELOW THAT OF IGNITION TEMPERATURE OF THE HAZARDOUS SUBSTANCE INVOLVED BEFORE BEING RELEASED INTO THE HAZARDOUS ATMOSPHERE.

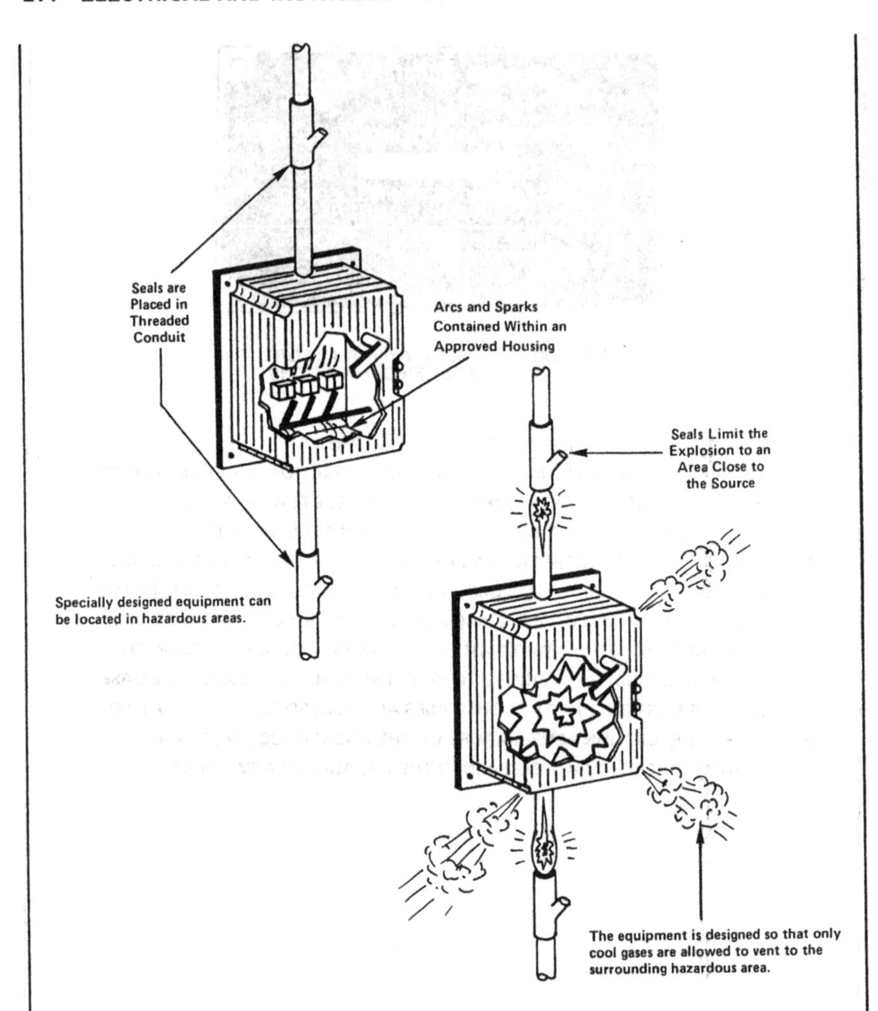

WHEN ARCS AND SPARKS CAUSE IGNITION OF FLAMMABLE GASES, VAPORS AND LIQUIDS, THE EQUIPMENT CONTAINS THE EXPLOSION AND VENTS ONLY COOL GASES INTO THE SURROUNDING HAZARDOUS AREA.

FIGURE 79. EXPLOSION OCCURRING IN APPROVED EQUIPMENT

WHEN AN INTERNAL EXPLOSION OCCURS, IT TENDS TO DISTORT THE SHAPE OF THE ENCLOSURE FROM RECTANGULAR TO ELLIPTICAL AS EXAGGERATED IN FIGURE 80. ADEQUATE STRENGTH IS ONE REQUIREMENT FOR THE DESIGN OF AN EXPLOSION-PROOF ENCLOSURE: A SAFETY FACTOR OF 4 IS GENERALLY USED. TO PREVENT FAILURE OF THE ENCLOSURE, OPENINGS ARE DESIGNED TO RELIEVE THE PRESSURE OF THE EXPANDING GASES. ALL JOINTS AND FLANGES ARE HELD TO NARROW TOLERANCES – THE ACCURATELY MACHINED JOINTS ACT TO COOL THE HOT GASES RESULTING FROM AN INTERNAL EXPLOSION SO THAT BY THE TIME THEY REACH THE OUTSIDE HAZARDOUS ATMOSPHERE THEY ARE TOO COOL TO CAUSE IGNITION.

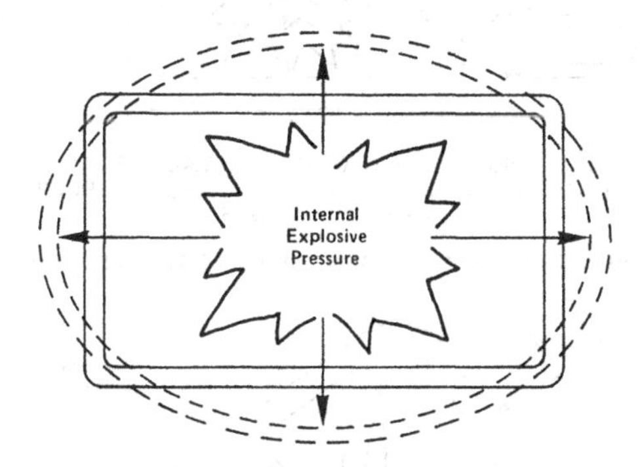

FIGURE 80. INTERNAL EXPLOSIVE PRESSURE

THERE ARE TWO COMMON ENCLOSURE DESIGNS: THREADED-JOINT ENCLOSURES (SEE FIGURE 81) AND GROUND-JOINT ENCLOSURES (SEE FIGURE 82). WHEN HOT GASES TRAVEL THROUGH THE VERY SMALL OPENINGS IN EITHER OF THESE JOINTS, THEY ARE COOLED BEFORE REACHING THE SURROUNDING HAZARDOUS ATMOSPHERE.

OTHER DESIGN REQUIREMENTS, SUCH AS SEALING, PREVENT THE PASSAGE OF GASES, VAPORS OR FUMES FROM ONE PORTION OF AN ELECTRICAL SYSTEM TO ANOTHER. MOTORS, WHICH TYPICALLY CONTAIN SPARKING BRUSHES OR COMMUTATORS AND TEND TO HEAT UP, MUST ALSO BE DESIGNED TO PROVIDE FOR THE CONTROL OF INTERNAL EXPLOSIONS.

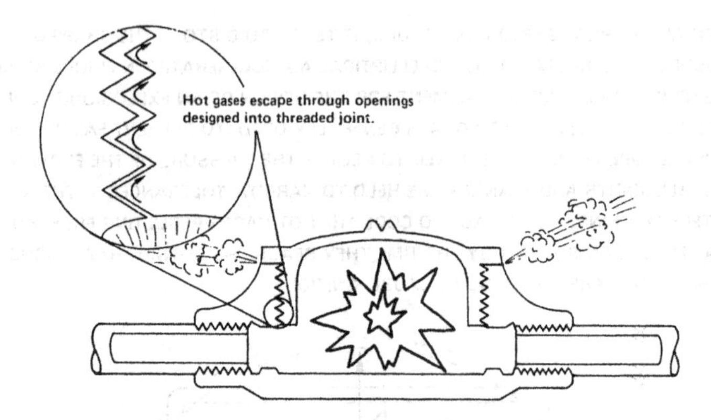

THE CASE OF THE ENCLOSURE IS MADE OF CAST METAL, STRONG ENOUGH TO WITHSTAND THE MAXIMUM EXPLOSION PRESSURE OF A SPECIFIC GROUP OF HAZARDOUS GASES OR VAPORS.

FIGURE 81. OPENINGS DESIGNED INTO THREADED JOINT

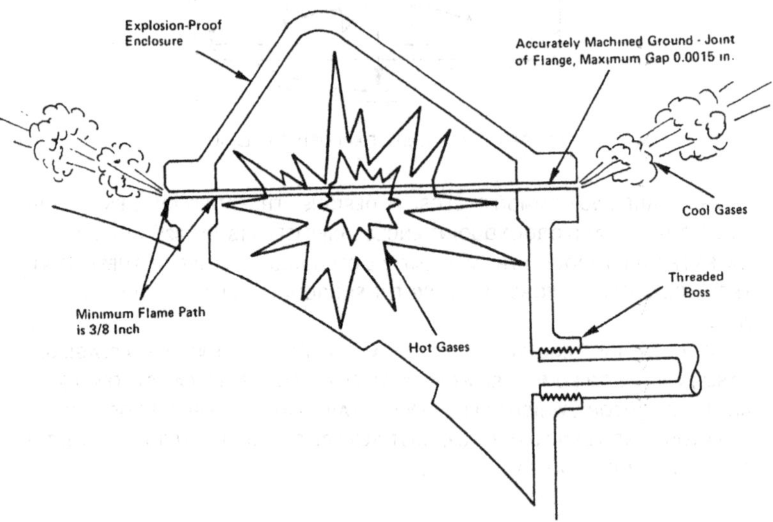

HOT BURNING GASES ARE COOLED AS THEY PASS THROUGH THE GROUND-JOINT OF FLANGES, DESIGNED WITHIN NARROW STANDARD TOLERANCES.

FIGURE 82. OPENINGS DESIGNED INTO GROUND JOINT

BECAUSE THE EXPLOSION CHARACTERISTICS OF HAZARDOUS SUBSTANCES VARY WITH THE SPECIFIC MATERIAL INVOLVED, EACH GROUP REQUIRES SPECIAL DESIGN CONSIDERATIONS. FOR CLASS I HAZARDOUS LOCATIONS, THERE ARE FOUR GROUPS – A, B, C, AND D. SEE TABLE 4A. DESIGN CHARACTERISTICS FOR THESE FOUR GROUPS REQUIRE THE CONTAINMENT OF MAXIMUM EXPLOSION PRESSURE, MAXIMUM SAFE CLEARANCE BETWEEN PARTS OF ENCLOSURES – INCLUDING THREADED JOINTS OR JOINTS THAT ARE GROUND TO NARROW TOLERANCES – AND OPERATION AT A TEMPERATURE BELOW THE IGNITION TEMPERATURE OF THE ATMOSPHERIC MIXTURE INVOLVED.

TABLE 4A. CHEMICALS BY GROUPS – CLASS I

GROUP A ATMOSPHERES:	ACETYLENE
GROUP B ATMOSPHERES:	ACROLEIN (INHIBITED), ARSINE, BUTADIENE, ETHYLENE OXIDE, HYDROGEN, MANUFACTURED GASES CONTAINING MORE THAN 30% HYDROGEN (BY VOLUME), PROPYLENE OXIDE, PROPYLNITRATE
GROUP C ATMOSPHERES:	ACETALDEHYDE, ALLYL ALCOHOL, N-BUTYRALDEHYDE, CARBON MONOXIDE, CROTONALDEHYDE, CYCLOPROPANE, DIETHYL ETHER, DIETHYLAMINE, EPICHLOROHYDRINE, ETHYLENE, ETHYLENIMINE, ETHYL MERCAPTAN, ETHYL SULFIDE, HYDROGEN CYANIDE, HYDROGEN SULFIDE, MORPHOLINE, 2-NITROPROPANE, TETRAHYDROFURAN, UNSYMMETRICAL DIMETHYL HYDRAZINE (UDMH 1, 1-DIMETHYL HYDRAZINE)
GROUP D ATMOSPHERES:	ACETIC ACID (GLACIAL), ACETONE, ACRYLONITRILE, AMMONIA, BENZENE, BUTANE, 1-BUTANOL (BUTYL ALCOHOL), 2-BUTANOL (SECONDARY BUTYL ALCOHOL), N-BUTYL ACETATE, ISOBUTYL ACETATE, DI-ISOBUTYLENE, ETHANE, ETHANOL (ETHYL ALCOHOL), ETHYL ACETATE, ETHYL ACRYLATE (INHIBITED), ETHYLENE DIAMINE (ANHYDROUS), ETHYLENE DICHLORIDE, ETHYLENE GLYCOL MONOMETHYL ETHER, GASOLINE, HEPTANES, HEXANES, ISOPRENE, ISOPROPYL ETHER, MESITYL OXIDE, METHANE (NATURAL GAS), METHANOL (METHYL ALCOHOL), 3-METHYL-1-BUTANOL (ISOAMYL ALCOHOL), METHYL ETHYL KETONE, METHYL ISOBUTYL KETONE, 2-METHYL-1-PROPANOL (ISOBUTYL ALCOHOL), 2-METHYL-2-PROPANOL (TERTIARY BUTYL ALCOHOL), PETROLEUM NAPHTHA, PYRIDINE, OCTANES, PENTANES, 1-PENTANOL (AMYL ALCOHOL), PROPANE, 1-PROPANOL (PROPYL ALCOHOL), 2-PROPANOL (ISOPROPYL ALCOHOL), PROPYLENE, STYRENE, TOLUENE, VINYL ACETATE, VINYL CHLORIDE, XYLENES

SOURCE: TABLE 500-2, ARTICLE 500 – HAZARDOUS (CLASSIFIED) LOCATIONS, 1981 NATIONAL ELECTRICAL CODE, NATIONAL FIRE PROTECTION ASSOCIATION, BOSTON, MASS.

DUST-IGNITION-PROOF EQUIPMENT

IN CLASS II, DIVISION 1 LOCATIONS, EQUIPMENT MUST GENERALLY BE DUST-IGNITION-PROOF. SECTION 502-1 OF THE NEC DEFINES DUST-IGNITION-PROOF AS EQUIPMENT "ENCLOSED IN A MANNER THAT WILL EXCLUDE IGNITIBLE AMOUNTS OF DUST OR AMOUNTS THAT MIGHT AFFECT PERFORMANCE OR RATING AND THAT, WHERE INSTALLED AND PROTECTED IN ACCORDANCE WITH THIS CODE, WILL NOT PERMIT ARCS, SPARKS, OR HEAT OTHERWISE GENERATED OR LIBERATED INSIDE THE ENCLOSURE TO CAUSE IGNITION OF EXTERIOR ACCUMULATIONS OR ATMOSPHERIC SUSPENSIONS OF A SPECIFIED DUST ON OR IN THE VICINITY OF THE ENCLOSURE."

DUST-IGNITION-PROOF EQUIPMENT IS DESIGNED TO KEEP IGNITIBLE AMOUNTS OF DUST FROM ENTERING THE ENCLOSURE. IN ADDITION, DUST MAY ACCUMULATE ON ELECTRICAL EQUIPMENT, CAUSING OVERHEATING OF THE EQUIPMENT, AS WELL AS THE DEHYDRATION OR GRADUAL CARBONIZATION OF ORGANIC DUST DEPOSITS. OVERHEATED EQUIPMENT MAY MALFUNCTION AND CAUSE A FIRE. DUST THAT HAS CARBONIZED IS SUSCEPTABLE TO SPONTANEOUS IGNITION OR SMOLDERING. THEREFORE, EQUIPMENT MUST ALSO BE DESIGNED TO OPERATE BELOW THE IGNITION TEMPERATURE OF THE SPECIFIC DUST INVOLVED EVEN WHEN BLANKETED. THE SHAPE OF THE ENCLOSURE MUST BE DESIGNED TO MINIMIZE DUST ACCUMULATION WHEN FIXTURES ARE OUT OF REACH OF NORMAL HOUSEKEEPING ACTIVITIES, E.G., LIGHTING FIXTURE CANOPYS.

IN CLASS II HAZARDOUS LOCATIONS THERE ARE THREE GROUPS – E, F, AND G. (SEE TABLE 4B.) SPECIAL DESIGNS ARE REQUIRED TO PREVENT DUST FROM ENTERING INTO THE ELECTRICAL EQUIPMENT ENCLOSURE. ASSEMBLY JOINTS AND MOTOR SHAFT OPENINGS MUST BE TIGHT ENOUGH TO PREVENT DUST FROM ENTERING THE ENCLOSURE. IN ADDITION THE DESIGN MUST TAKE INTO ACCOUNT THE INSULATING EFFECTS OF DUST LAYERS ON EQUIPMENT AND MUST ENSURE THAT THE EQUIPMENT WILL OPERATE BELOW THE IGNITION TEMPERATURE OF THE DUST INVOLVED. IF CONDUCTIVE COMBUSTIBLE DUSTS ARE PRESENT, THE DESIGN OF EQUIPMENT MUST TAKE THE SPECIAL NATURE OF THESE DUSTS INTO ACCOUNT.

IN GENERAL, EQUIPMENT WHICH IS APPROVED EXPLOSION-PROOF IS NOT DESIGNED FOR, AND IS NOT ACCEPTABLE FOR USE IN, CLASS II LOCATIONS, UNLESS SPECIFICALLY APPROVED FOR USE IN SUCH LOCATIONS. FOR EXAMPLE, SINCE GRAIN DUST HAS A LOWER IGNITION TEMPERATURE THAN THAT OF MANY FLAMMABLE VAPORS, EQUIPMENT APPROVED FOR CLASS I LOCATIONS MAY OPERATE AT A TEMPERATURE THAT IS TOO HIGH FOR CLASS II LOCATIONS. ON THE OTHER HAND, EQUIPMENT THAT IS DUST-IGNITION-PROOF IS GENERALLY ACCEPTABLE FOR USE IN CLASS III LOCATIONS, SINCE THE SAME DESIGN CONSIDERATIONS ARE INVOLVED.

TABLE 4B. CHEMICALS BY GROUPS – CLASS II

GROUP E ATMOSPHERES:	METAL DUST, INCLUDING ALUMINUM, MAGNESIUM, AND THEIR COMMERCIAL ALLOYS, AND OTHER METALS OF SIMILARLY HAZARDOUS CHARACTERISTICS HAVING RESISTIVITY OF 10^2 OHM-CENTIMETER OR LESS.
GROUP F ATMOSPHERES:	CARBON BLACK, CHARCOAL, COAL, OR COKE DUSTS
GROUP G ATMOSPHERES:	FLOUR, STARCH, GRAIN DUST, OR COMBUSTIBLE PLASTIC OR CHEMICAL DUSTS HAVING RESISTIVITY GREATER THAN 10^8 OHM-CENTIMETER.

MARKING

APPROVED EQUIPMENT MUST BE MARKED TO INDICATE THE CLASS, GROUP, AND OPERATING TEMPERATURE RANGE (BASED ON A 40° C AMBIENT TEMPERATURE) IN WHICH IT IS DESIGNED TO BE USED. FURTHERMORE, THE TEMPERATURE MARKED ON THE EQUIPMENT MUST NOT BE GREATER THAN THE IGNITION TEMPERATURE OF THE SPECIFIC GASES OR VAPORS IN THE AREA.

THERE ARE, HOWEVER, FOUR EXCEPTIONS TO THIS MARKING REQUIREMENT. FIRST, EQUIPMENT THAT DOES NOT PRODUCE HEAT (FOR EXAMPLE, JUNCTION BOXES OR CONDUITS) AND EQUIPMENT THAT DOES PRODUCE HEAT BUT THAT HAS A MAXIMUM SURFACE TEMPERATURE OF LESS THAN 100°C (OR 212°F) ARE NOT REQUIRED TO BE MARKED WITH THE OPERATING TEMPERATURE RANGE. THE HEAT NORMALLY RELEASED FROM THIS EQUIPMENT CANNOT IGNITE GASES, LIQUIDS, VAPORS, OR DUSTS.

SECOND, ANY PERMANENT LIGHTING FIXTURES THAT ARE APPROVED AND MARKED FOR USE IN CLASS I, DIVISION 2 LOCATIONS DO NOT NEED TO BE MARKED TO SHOW A SPECIFIC GROUP. THIS IS BECAUSE THESE FIXTURES ARE ACCEPTABLE FOR USE WITH ALL OF THE CHEMICAL GROUPS FOR CLASS I (THAT IS, GROUPS A, B, C, AND D).

THIRD, FIXED GENERAL-PURPOSE EQUIPMENT IN CLASS I LOCATIONS, OTHER THAN LIGHTING FIXTURES, THAT IS ACCEPTABLE FOR USE IN DIVISION 2 LOCATIONS DOES NOT HAVE TO BE LABELED ACCORDING TO CLASS, GROUP, DIVISION, OR OPERATING TEMPERATURE. THIS TYPE OF EQUIPMENT DOES NOT CONTAIN ANY DEVICES THAT MIGHT PRODUCE ARCS OR SPARKS AND, THEREFORE, IS NOT A POTENTIAL IGNITION SOURCE. FOR EXAMPLE, SQUIRREL-CAGE INDUCTION MOTORS WITHOUT BRUSHES, SWITCHING MECHANISMS OR SIMILAR ARC-PRODUCING DEVICES ARE PERMITTED IN CLASS I, DIVISION 2 LOCATIONS (SEE NEC SECTION 501-8(b)); THEREFORE, THEY NEED NO MARKING.

FOURTH, FOR CLASS II, DIVISION 2 AND CLASS III LOCATIONS, FIXED DUST-TIGHT EQUIPMENT (OTHER THAN LIGHTING FIXTURES) IS NOT REQUIRED TO BE MARKED. IN THESE LOCATIONS, DUST-TIGHT EQUIPMENT DOES NOT PRESENT A HAZARD SO IT NEED NOT BE IDENTIFIED.

* SAFE FOR THE HAZARDOUS (CLASSIFIED) LOCATION

UNDER THIS OPTION, EQUIPMENT INSTALLED IN HAZARDOUS (CLASSIFIED) LOCATIONS MUST BE OF A TYPE AND DESIGN WHICH PROVIDES PROTECTION FROM THE HAZARDS ARISING FROM THE COMBUSTIBILITY AND FLAMMABILITY OF VAPORS, LIQUIDS, GASES, DUSTS, OR FIBERS. THE EMPLOYER HAS THE RESPONSIBILITY OF DEMONSTRATING THAT THE INSTALLATION MEETS THIS REQUIREMENT. GUIDELINES FOR INSTALLING EQUIPMENT UNDER THIS OPTION ARE CONTAINED IN THE NATIONAL ELECTRICAL CODE IN EFFECT AT THE TIME OF INSTALLATION OF THAT EQUIPMENT. COMPLIANCE WITH THESE GUIDELINES ARE NOT THE ONLY MEANS OF COMPLYING WITH THIS OPTION; HOWEVER, THE EMPLOYER MUST DEMONSTRATE THAT HIS INSTALLATION IS SAFE FOR THE HAZARDOUS (CLASSIFIED) LOCATION.

THE FOLLOWING PARAGRAPHS SUMMARIZE INSTALLATION PRACTICES GIVEN IN THE 1981 NEC. THESE PRACTICES WOULD BE AN ACCEPTABLE MEANS OF COMPLYING WITH THIS THIRD OPTION GIVEN FOR EQUIPMENT IN HAZARDOUS LOCATIONS.

- CLASS I, DIVISION 1

ARTICLE 501 OF THE NATIONAL ELECTRICAL CODE (NEC) CONTAINS INSTALLATION REQUIREMENTS FOR ELECTRICAL WIRING AND EQUIPMENT USED IN CLASS I HAZARDOUS AREAS. THE REQUIREMENTS AS THEY PERTAIN TO CLASS I DIVISION 1 HAZARDOUS LOCATIONS ARE SUMMARIZED IN FIGURE 83 AND TABLE 5. THE REQUIREMENTS FOR CLASS I DIVISION 2 LOCATIONS ARE SUMMARIZED IN FIGURE 88 AND TABLE 6.

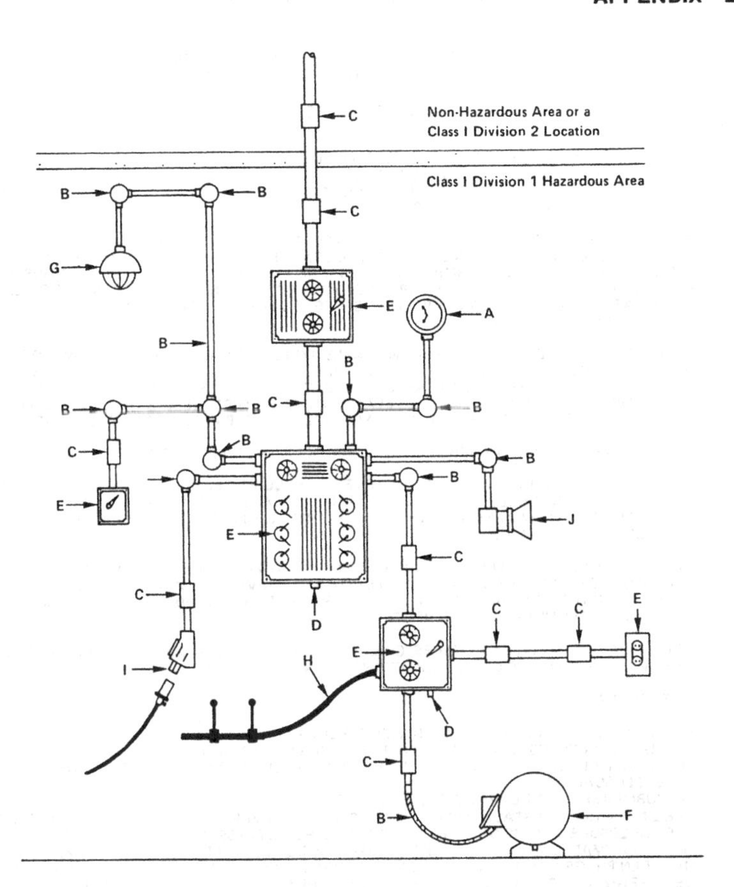

FIGURE 83. CLASS I DIVISION 1 HAZARDOUS LOCATION
(SEE TABLE 5)

TABLE 5. SUMMARY OF EQUIPMENT REQUIREMENTS FOR CLASS I DIVISION 1 HAZARDOUS LOCATIONS

(SEE FIGURE 83)

A. METERS, RELAYS, AND INSTRUMENTS, SUCH AS VOLTAGE OR CURRENT METERS AND PRESSURE OR TEMPERATURE SENSORS, MUST BE IN ENCLOSURES APPROVED FOR CLASS I, DIVISION 1 LOCATIONS. SUCH ENCLOSURES INCLUDE EXPLOSION-PROOF AND PURGED AND PRESSURIZED ENCLOSURES. SEE NEC SECTION 501-3(a).

B. WIRING METHODS ACCEPTABLE FOR USE IN CLASS I DIVISION 1 LOCATION INCLUDE: THREADED RIGID METAL OR STEEL INTERMEDIATE METAL CONDUIT AND TYPE MI CABLE. FLEXIBLE FITTINGS, SUCH AS MOTOR TERMINATIONS, MUST BE APPROVED FOR CLASS I LOCATIONS. ALL BOXES AND ENCLOSURES MUST BE EXPLOSION-PROOF AND THREADED FOR CONDUIT OR CABLE TERMINATIONS. ALL JOINTS MUST BE WRENCH TIGHT WITH A MINIMUM OF FIVE THREADS ENGAGED. SEE NEC 501-4(a).

C. SEALING IS REQUIRED FOR CONDUIT AND CABLE SYSTEMS TO PREVENT THE PASSAGE OF GASES, VAPORS, AND FLAME FROM ONE PART OF THE ELECTRICAL INSTALLATION TO ANOTHER THROUGH THE CONDUIT. TYPE MI CABLE INHERENTLY PREVENTS THIS FROM HAPPENING BY ITS CONSTRUCTION; HOWEVER, IT MUST BE SEALED TO KEEP MOISTURE AND OTHER FLUIDS FROM ENTERING THE CABLE AT TERMINATIONS. SEE FIGURE 84. SEE ALSO NEC SECTION 501-5.

(1) SEALS ARE REQUIRED WHERE CONDUIT PASSES FROM DIVISION 1 TO DIVISION 2 OR NON-HAZARDOUS LOCATIONS.
(2) SEALS ARE REQUIRED WITHIN 18 INCHES FROM ENCLOSURES CONTAINING ARCING DEVICES.
(3) SEALS ARE REQUIRED IF CONDUIT IS 2 INCHES IN DIAMETER OR LARGER ENTERING AN ENCLOSURE CONTAINING TERMINATIONS, SPLICES, OR TAPS. SEE FIGURE 85 FOR A DESCRIPTION OF SEALS.

D. DRAINAGE IS REQUIRED WHERE LIQUID OR CONDENSED VAPOR MAY BE TRAPPED WITHIN AN ENCLOSURE OR RACEWAY. AN APPROVED SYSTEM OF PREVENTING ACCUMULATIONS OR TO PERMIT PERIODIC DRAINAGE ARE TWO METHODS TO CONTROL CONDENSATION OF VAPORS AND LIQUID ACCUMULATION. SEE NEC SECTION 501-5(f).

E. ARCING DEVICES, SUCH AS SWITCHES, CIRCUIT BREAKERS, MOTOR CONTROLLERS, AND FUSES, MUST BE APPROVED FOR CLASS I LOCATIONS. SEE NEC SECTION 501-6(A).

F. MOTORS SHALL BE

(1) APPROVED FOR USE IN CLASS I, DIVISION 1 LOCATIONS;
(2) TOTALLY ENCLOSED WITH POSITIVE PRESSURE VENTILATION;
(3) TOTALLY ENCLOSED INERT-GAS-FILLED WITH A POSITIVE PRESSURE WITHIN THE ENCLOSURE; OR
(4) SUBMERGED IN A FLAMMABLE LIQUID OR GAS.

THE LAST KIND OF INSTALLATION IS PERMISSIBLE, HOWEVER, ONLY WHEN THERE IS PRESSURE ON THE ENCLOSURE THAT IS GREATER THAN ATMOSPHERIC PRESSURE AND THE LIQUID OR GAS IS ONLY FLAMMABLE IN AIR. THIS TYPE OF MOTOR IS NOT PERMITTED TO BE ENERGIZED UNTIL IT HAS BEEN PURGED OF ALL AIR. THE LATTER THREE TYPES OF MOTORS MUST BE ARRANGED TO BE DE-ENERGIZED SHOULD THE PRESSURE FAIL OR THE SUPPLY OF LIQUID OR GAS FAIL – AS WITH THE SUBMERGED TYPE. TYPES (2) AND (3) MAY NOT OPERATE AT A SURFACE TEMPERATURE ABOVE 80 PERCENT OF THE IGNITION TEMPERATURE OF THE GAS OR VAPOR INVOLVED. SEE NEC SECTION 501-8(a).

G. LIGHTING FIXTURES, BOTH FIXED AND PORTABLE, MUST BE EXPLOSION-PROOF AND GUARDED AGAINST PHYSICAL DAMAGE. SEE NEC SECTION 501-9(a).

H. FLEXIBLE CORDS MUST BE DESIGNED FOR EXTRA HARD USAGE, CONTAIN AN EQUIPMENT GROUNDING CONDUCTOR (SEE FIGURE 83), BE SUPPORTED SO THAT THERE WILL BE NO TENSION ON THE TERMINAL CONNECTIONS, AND BE PROVIDED WITH SEALS WHERE THEY ENTER EXPLOSION-PROOF ENCLOSURES. SEE NEC SECTION 501-11.

NOTE: NEC-NATIONAL ELECTRICAL CODE, NFPA 70.

TABLE 5 (CONTINUED)

I. RECEPTACLES AND ATTACHMENT PLUGS FOR USE WITH PORTABLE EQUIPMENT MUST BE APPROVED EXPLOSION-PROOF AND PROVIDED WITH AN EQUIPMENT GROUNDING CONNECTION. SEE NEC SECTION 501-12.

J. SIGNALING, ALARM, REMOTE CONTROL AND COMMUNICATIONS SYSTEMS ARE REQUIRED TO BE APPROVED FOR CLASS I, DIVISION 1 LOCATIONS REGARDLESS OF VOLTAGE. SEE NEC SECTION 501-14(A).

K. EQUIPMENT GROUNDING IS REQUIRED OF ALL NON-CURRENT-CARRYING METAL PARTS OF THE ELECTRICAL SYSTEM. IN ADDITION, LOCK NUTS AND BRUSHINGS MUST NOT BE RELIED UPON FOR ELECTRICAL CONNECTION BETWEEN RACEWAYS AND EQUIPMENT. IF LOCKNUTS AND BUSHINGS ARE USED, BONDING JUMPERS ARE REQUIRED. SEE NEC SECTION 501-16.

NOTE: NEC-NATIONAL ELECTRICAL CODE, NFPA 70.

THE FOLLOWING SECTIONS FURTHER EXPLAIN REQUIREMENTS FOR TYPE MI CABLE, SEALING, MOTORS, GROUNDING, AND BONDING.

TYPE MI (MINERAL INSULATED) CABLE

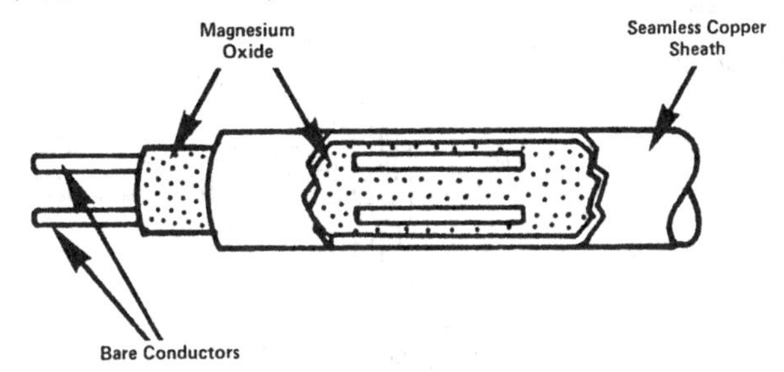

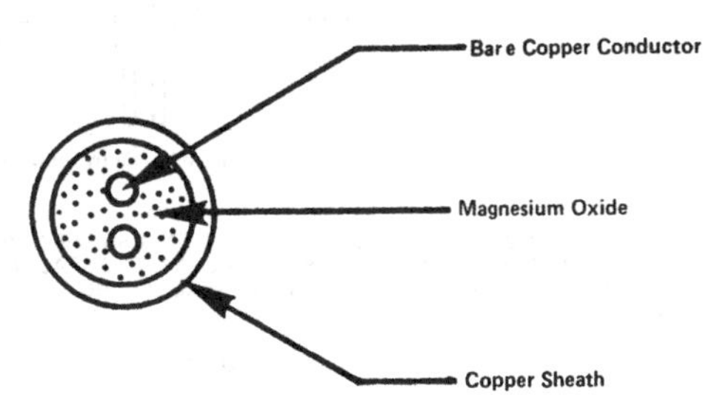

TYPE MI CABLE IS A MINERAL-INSULATED CABLE OF COPPER CONDUCTORS IN TIGHTLY COMPRESSED MAGNESIUM OXIDE THAT IS ENCLOSED IN A LIQUIDTIGHT AND GASTIGHT COPPER COVERING. SINCE TYPE MI CABLE FITTINGS SUITABLE FOR NONHAZARDOUS LOCATIONS MAY NOT BE EXPLOSION PROOF, FITTINGS USED WITH THE CABLE MUST BE SPECIALLY DESIGNED FOR CLASS I LOCATIONS. BOXES, FITTINGS, AND JOINTS USED WITH THE CABLE MUST ALSO BE EXPLOSION-PROOF.

FIGURE 84. CONSTRUCTION OF TYPE MI (MINERAL INSULATED) CABLE

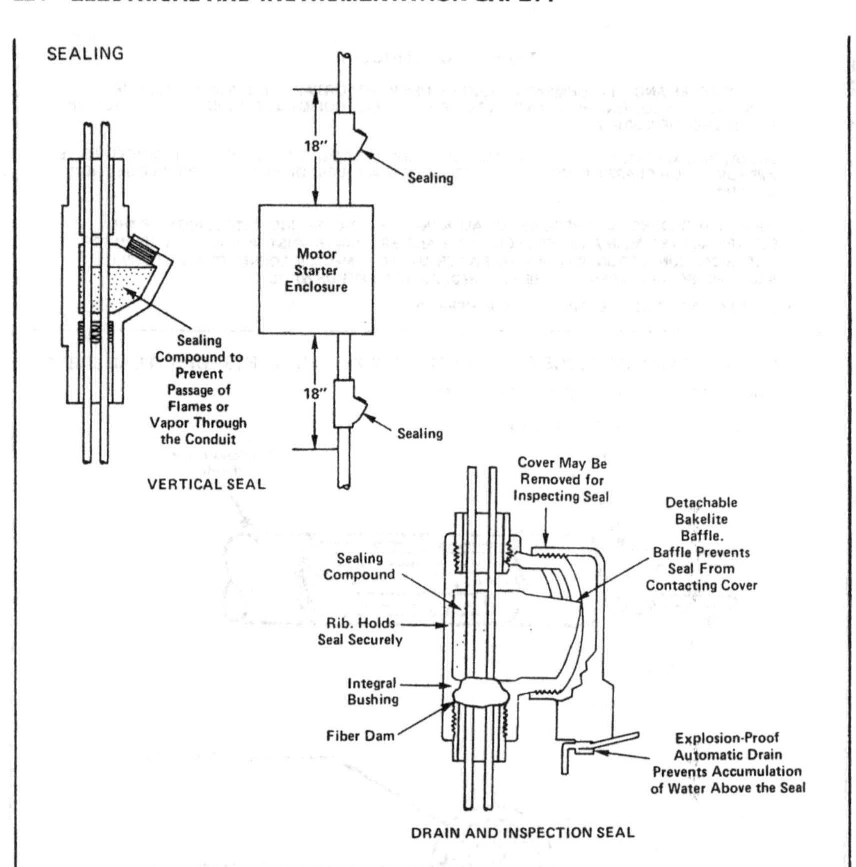

A FIBER DAM IS PLACED IN THE CONDUIT WHERE IT ENTERS THE SEAL FITTING. THIS SERVES TO CONTAIN THE SEALING COMPOUND WHILE IT HARDENS. THE SEALING COMPOUND FORMS A TIGHT SEAL TO PREVENT THE PASSAGE OF VAPORS OR FLAMES THROUGH THE CONDUIT SYSTEM. SEE NEC SECTION 501-5.

FIGURE 85. SEALING

MOTORS

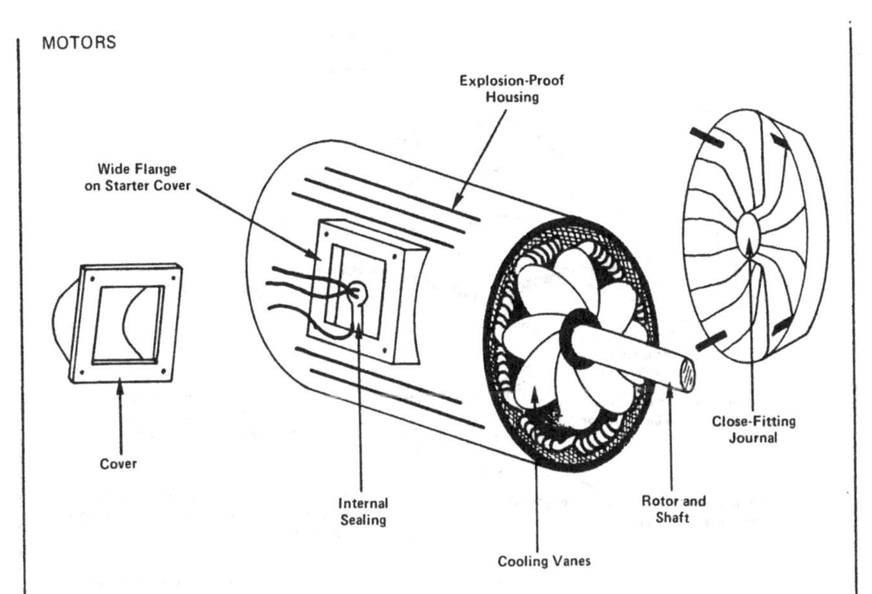

FIGURE 86 SHOWS AN INTERNAL VIEW OF A TOTALLY ENCLOSED FAN-COOLED EXPLOSION-PROOF MOTOR. THE ROTOR AND ITS WINDINGS AND FAN ARE COMPLETELY ENCLOSED. AN INTERNAL SEAL, A WIDE FLANGE ON THE STARTER COVER, AND A CLOSE-FITTING JOURNAL PREVENT THE ESCAPE OF HOT GASES OR FLAMES FROM THE ENCLOSURE. AN INTERNAL FAN CIRCULATES AIR INSIDE THE ENCLOSURE, TRANSFERRING THE HEAT FROM THE WINDINGS TO THE ENCLOSURE. THE FRAME OR ENCLOSURE IS EXPLOSION-PROOF AND MAY HAVE AN EXTERNAL FAN THAT FORCES AIR OVER ITS OUTSIDE SURFACES. THIS FORCED EXTERNAL CIRCULATION WILL PROVIDE MORE EFFECTIVE MOTOR COOLING THAN WOULD NATURAL AIR CIRCULATION. HOWEVER, NONE OF THIS EXTERNAL AIR COMES IN CONTACT WITH THE WINDINGS.

FIGURE 86. DESIGN FEATURES OF A TOTALLY ENCLOSED, FAN-COOLED, EXPLOSION-PROOF MOTOR

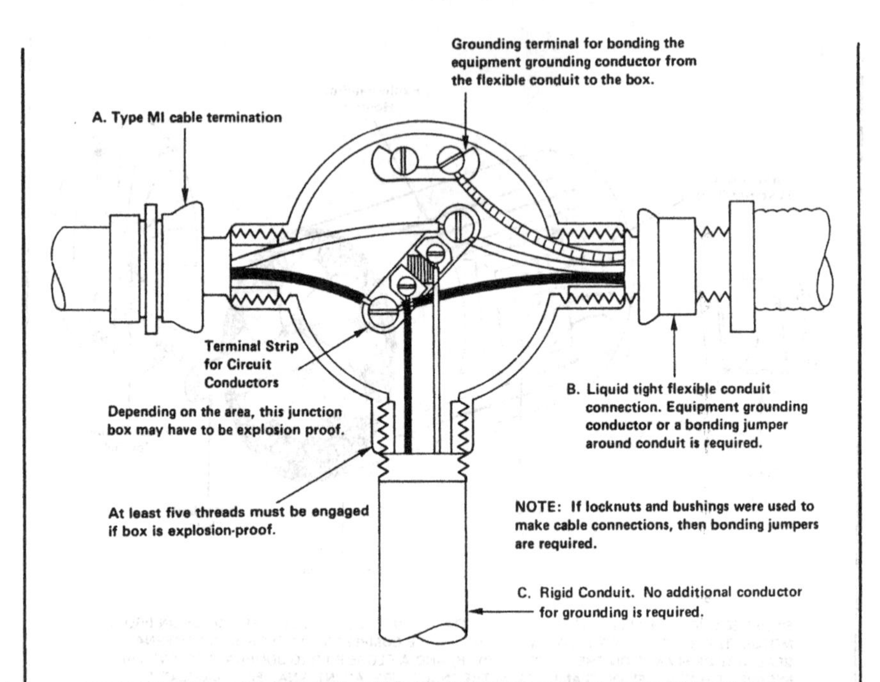

IN CLASS I DIVISION 1 AND DIVISION 2 HAZARDOUS LOCATIONS, EXPOSED NON-CURRENT-CARRYING METAL PARTS OF EQUIPMENT, SUCH AS FRAMES AND CABINETS, MUST BE GROUNDED. THE GROUND MUST PROVIDE A PATH BACK TO THE SOURCE IF AN ACCIDENTAL FAULT OCCURS.

BONDING IS ALSO REQUIRED TO PROVIDE A PERMANENT GROUND FOR EXPOSED METAL PARTS. TO BE CONSIDERED EFFECTIVE, BONDING MUST PREVENT THE OCCURRENCE OF ARCS OR SPARKS CAUSED BY POOR CONNECTIONS. FIGURE 87 SHOWS A TYPICAL GROUNDING AND BONDING TECHNIQUE. SEE NEC ARTICLE 250 AND SECTION 501-16.

SPECIAL CARE MUST BE TAKEN TO MAKE PROPER BONDING CONNECTIONS NOT ONLY TO ASSURE THAT THERE IS A CONTINUOUS EQUIPMENT GROUNDING PATH BUT TO BE POSITIVE THAT NO ARCING OR SPARKING WILL TAKE PLACE BETWEEN CONNECTIONS. LOCK NUT-BUSHINGS AND DOUBLE-LOCK NUT CONNECTORS CANNOT BE RELIED UPON FOR BONDING PURPOSES. FIGURE 87 ILLUSTRATES THREE TYPICAL ARRANGEMENTS OF CONDUIT AND CABLE CONNECTIONS TO AN EXPLOSION-PROOF ENCLOSURE, AND BONDING METHODS. THESE ARE: A) TYPE MI CABLE TERMINATION, B) FLEXIBLE CONDUIT CONNECTION, AND C) RIGID CONDUIT CONNECTION.

FIGURE 87. BONDING IN CLASS I HAZARDOUS (CLASSIFIED) LOCATIONS

CLASS 1 DIVISION 2

THE REQUIREMENTS FOR CLASS I DIVISION 2 HAZARDOUS LOCATIONS ARE SUMMARIZED IN FIGURE 88 AND TABLE 6 AS FOLLOWS:

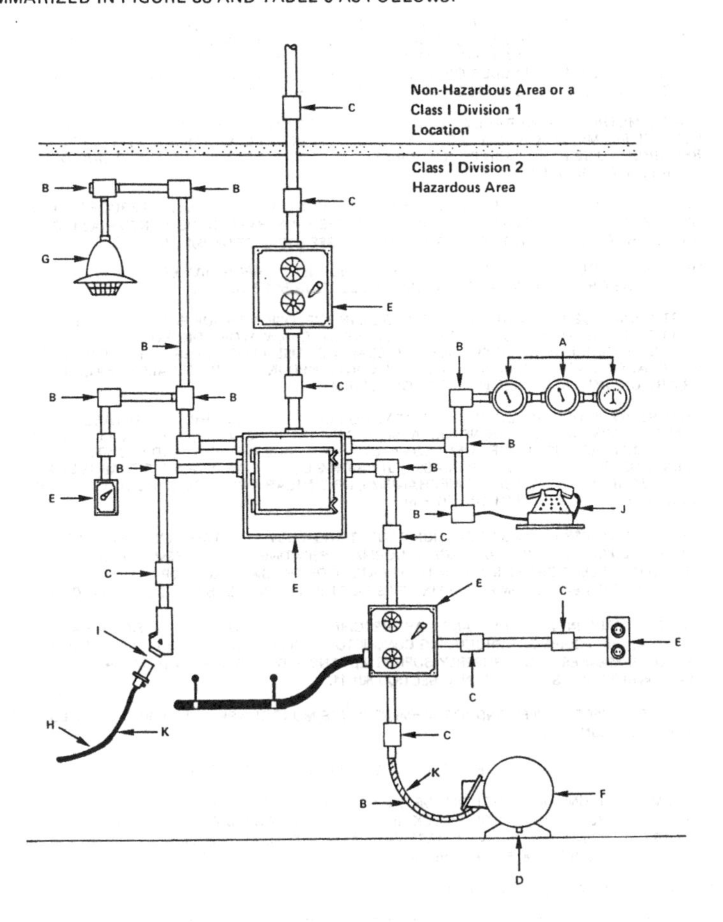

FIGURE 88. CLASS I DIVISION 2 HAZARDOUS LOCATIONS (SEE TABLE 6)

TABLE 6. SUMMARY OF CLASS 1, DIVISION 2 HAZARDOUS LOCATIONS

(SEE FIGURE 88)

A. METERS, INSTRUMENTS AND RELAYS IN CLASS 1, DIVISION 2 LOCATIONS MUST BE IN APPROVED EXPLOSION-PROOF ENCLOSURES. HOWEVER, GENERAL-PURPOSE EQUIPMENT MAY BE USED, IF CIRCUIT INTERRUPTING CONTACTS ARE IMMERSED IN OIL OR ENCLOSED IN A HERMETICALLY SEALED CHAMBER OR IN CIRCUITS THAT DO NOT RELEASE ENOUGH ENERGY TO IGNITE THE HAZARDOUS ATMOSPHERE. SEE NEC SECTION 501-3(b).

B. WIRING METHODS: GENERALLY, THREADED RIGID OR INTERMEDIATE CONDUIT OR TYPES PLTC, MI, MC, MV, TC, OR SNM CABLE SYSTEMS MUST BE USED. BOXES AND FITTINGS ARE NOT REQUIRED TO BE EXPLOSION PROOF UNLESS THEY ENCLOSE ARCING OR SPARKING DEVICES. SEE NEC SECTION 501-4(b).

C. SEALS ARE REQUIRED FOR ALL CONDUIT SYSTEMS CONNECTED TO EXPLOSION-PROOF ENCLOSURES. SEALS ARE ALSO REQUIRED WHERE CONDUIT PASSES FROM HAZARDOUS TO NON–HAZARDOUS AREAS OR FROM DIVISION 1 TO DIVISON 2 AREAS. (SEE NEC SECTION 501-5(b).

D. DRAINAGE IS REQUIRED WHERE LIQUID OR CONDENSED VAPOR MAY BE TRAPPED WITHIN AN ENCLOSURE OR ALONG A RACEWAY. SEE NEC SECTION 501-5(f).

E. MOST ARCING DEVICES ARE REQUIRED TO BE IN EXPLOSION-PROOF ENCLOSURES. THESE INCLUDE ITEMS SUCH AS SWITCHES, CIRCUIT BREAKERS, MOTOR CONTROLLERS AND FUSES. HOWEVER, GENERAL PURPOSE ENCLOSURES MAY BE USED FOR CLASS 1, DIVISION 2 LOCATIONS. IF THE ARCING AND SPARKING PARTS ARE CONTAINED IN A HERMETICALLY SEALED CHAMBER OR ARE OIL IMMERSED. SEE NEC SECTION 501-6(b).

F. MOTORS, GENERATORS AND OTHER ROTATING ELECTRICAL MACHINERY SUITABLE FOR USE IN CLASS 1, DIVISION 1 LOCATIONS ARE ALSO ACCEPTABLE IN CLASS I, DIVISION 2 LOCATIONS. OTHER MOTORS MUST HAVE THEIR CONTACTS, SWITCHING DEVICES, AND RESISTANCE DEVICES IN ENCLOSURES SUITABLE FOR CLASS I, DIVISION 2 LOCATIONS (SEE NOTE E, ABOVE). MOTORS WITHOUT BRUSHES, SWITCHING MECHANISMS, OR SIMILAR ARC-PRODUCING DEVICES ARE ALSO ACCEPTABLE. SEE NEC SECTION 501-8(b).

G. LIGHTING FIXTURES IN CLASS 1 DIVISION 2 LOCATIONS MUST BE TOTALLY ENCLOSED AND PROTECTED FROM PHYSICAL DAMAGE. IF NORMAL OPERATING SURFACE TEMPERATURES EXCEED 80 PERCENT OF THE IGNITION TEMPERATURE OF THE GAS, LIQUID OR VAPOR INVOLVED, THEN EXPLOSION-PROOF FIXTURES MUST BE INSTALLED. SEE NEC SECTION 501-9(b).

H. FLEXIBLE CORDS IN DIVISIONS 1 AND 2 ARE REQUIRED TO: 1) BE SUITABLE FOR EXTRA HARD USAGE, 2) CONTAIN AN EQUIPMENT GROUNDING CONDUCTOR, 3) BE CONNECTED TO TERMINALS IN AN APPROVED MANNER, 4) BE PROPERLY SUPPORTED, AND 5) BE PROVIDED WITH SUITABLE SEALS WHERE NECESSARY. SEE NEC SECTION 501-11.

I. IN GENERAL, RECEPTACLES AND ATTACHMENT PLUGS MUST BE APPROVED FOR CLASS 1 LOCATIONS. SEE NEC SECTION 501-12.

J. SIGNALING SYSTEMS AND OTHER SIMILAR SYSTEMS: SEE NEC SECTION 501-14.

K. EQUIPMENT GROUNDING IS REQUIRED OF ALL NON-CURRENT-CARRYING METAL PARTS OF THE ELECTRICAL SYSTEM. IN ADDITION, LOCK NUTS AND BUSHINGS MUST NOT BE RELIED UPON FOR ELECTRICAL CONNECTION BETWEEN RACEWAYS AND EQUIPMENT. IF LOCKNUTS AND BUSHINGS ARE USED, BONDING JUMPERS ARE REQUIRED. SEE NEC SECTION 501-16.

NOTE: NEC-NATIONAL ELECTRICAL CODE, NFPA 70.

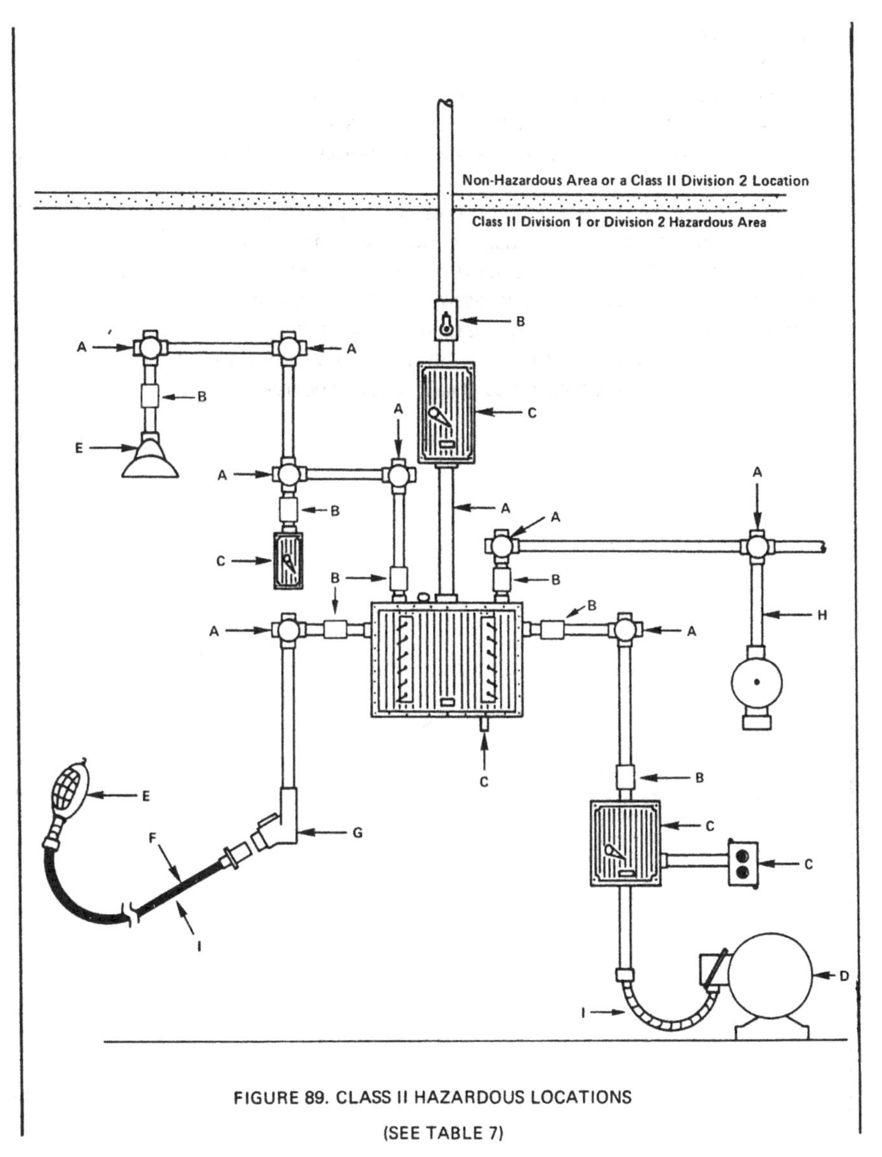

FIGURE 89. CLASS II HAZARDOUS LOCATIONS

(SEE TABLE 7)

CLASS II HAZARDOUS LOCATIONS

ARTICLE 502 OF THE NATIONAL ELECTRICAL CODE (NEC) IS CONCERNED WITH THE INSTALLATION REQUIREMENTS FOR ELECTRICAL WIRING AND EQUIPMENT USED IN CLASS II HAZARDOUS AREAS. THE REQUIREMENTS AS THEY PERTAIN TO CLASS II DIVISION 1 AND DIVISION 2 LOCATIONS ARE SUMMARIZED IN FIGURE 89 AND TABLE 7.

CLASS II LOCATIONS ARE HAZARDOUS BECAUSE OF THE PRESENCE OF COMBUSTIBLE DUST. AS DISCUSSED PREVIOUSLY, THESE DUSTS ARE BROKEN DOWN INTO THREE GROUPS – E, F, AND G. THE DUSTS ARE ALSO DIVIDED INTO TWO CATEGORIES: CONDUCTIVE (HAVING RESISTIVITY LESS THAN 10^5 OHM-CENTIMETER) AND NONCONDUCTIVE. WHERE CONDUCTIVE DUSTS ARE PRESENT, THERE ARE ONLY CLASS II, DIVISION 1 LOCATIONS. GROUP E DUSTS ARE CONDUCTIVE, SOME GROUP F DUSTS ARE CONDUCTIVE, AND GROUP G DUSTS ARE NONCONDUCTIVE.

TABLE 7. SUMMARY OF CLASS II HAZARDOUS LOCATIONS

(SEE FIGURE 89)

A. WIRING METHODS FOR CLASS II, DIVISION I LOCATIONS: BOXES AND FITTINGS CONTAINING ARCING AND SPARKING PARTS ARE REQUIRED TO BE IN DUST-IGNITION-PROOF ENCLOSURES. FOR OTHER THAN FLEXIBLE CONNECTIONS THREADED METAL CONDUIT OR TYPE MI CABLE WITH APPROVED TERMINATIONS IS REQUIRED FOR CLASS II, DIVISION 1 LOCATIONS. SEE NEC SECTION 502-4(a).

IN CLASS II DIVISION 2 LOCATIONS, BOXES AND FITTINGS ARE NOT REQUIRED TO BE DUST-IGNITION PROOF BUT MUST BE DESIGNED TO MINIMIZE THE ENTRANCE OF DUST AND PREVENT THE ESCAPE OF SPARKS OR BURNING MATERIAL. IN ADDITION TO THE WIRING SYSTEMS SUITABLE FOR DIVISION 1 LOCATIONS, THE FOLLOWING SYSTEMS ARE SUITABLE FOR DIVISION 2 LOCATIONS: ELECTRICAL METALLIC TUBING, DUST-TIGHT WIREWAYS, AND TYPES MC AND SNM CABLES. SEE NEC SECTION 502-4(b).

B. SUITABLE MEANS OF PREVENTING THE ENTRANCE OF DUST INTO A DUST-IGNITION-PROOF ENCLOSURE MUST BE PROVIDED WHERE A RACEWAY PROVIDES A PATH TO THE DUST-IGNITION-PROOF ENCLOSURE FROM ANOTHER ENCLOSURE THAT COULD ALLOW THE ENTRANCE OF DUST. SEE FIGURES 90-92. ALSO SEE NEC SECTION 502-5.

C. SWITCHES, CIRCUIT BREAKERS, MOTOR CONTROLLERS, AND FUSES INSTALLED IN CLASS II, DIVISION 1 LOCATIONS MUST BE DUST-IGNITION PROOF.

IN CLASS II, DIVISION 2 AREAS, ENCLOSURES FOR FUSES, SWITCHES, CURCUIT BREAKERS, AND MOTOR CONTROLLERS MUST BE DUST-TIGHT. SEE NEC SECTION 502-6.

D. IN CLASS II, DIVISION 1 LOCATIONS, MOTORS, GENERATORS, AND OTHER ROTATING ELECTRICAL MACHINERY MUST BE DUST-IGNITION PROOF OR TOTALLY ENCLOSED PIPE VENTILATED.

IN CLASS II, DIVISION 2 AREAS, ROTATING EQUIPMENT MUST BE ONE OF THE FOLLOWING TYPES:
1) DUST-IGNITION-PROOF,
2) TOTALLY ENCLOSED PIPE VENTILATED,
3) TOTALLY ENCLOSED NONVENTILATED, OR
4) TOTALLY ENCLOSED FAN COOLED.
UNDER CERTAIN CONDITIONS, STANDARD OPEN-TYPE MACHINES AND SELF-CLEANING SQUIRREL-CAGE MOTORS MAY BE USED. SEE NEC SECTION 502-8.

E. IN CLASS II, DIVISION 1 LOCATIONS, LIGHTING FIXTURES MUST BE DUST-IGNITION PROOF.

LIGHTING FIXTURES IN CLASS II DIVISION 2 LOCATIONS MUST BE DESIGNED TO MINIMIZE ACCUMULATION OF DUST AND MUST BE ENCLOSED TO PREVENT THE RELEASE OF SPARKS OR HOT METAL.

IN BOTH DIVISIONS, EACH FIXTURE MUST BE CLEARLY MARKED FOR THE MAXIMUM WATTAGE OF THE LAMP, SO THAT THE MAXIMUM PERMISSIBLE SURFACE TEMPERATURE FOR THE FIXTURE IS NOT EXCEEDED. ADDITIONALLY, FIXTURES MUST BE PROTECTED FROM DAMAGE. SEE NEC SECTION 502-11.

F. FLEXIBLE CORDS IN DIVISIONS 1 AND 2 ARE REQUIRED TO: 1) BE SUITABLE FOR EXTRA HARD USAGE, 2) CONTAIN AN EQUIPMENT GROUNDING CONDUCTOR, 3) BE CONNECTED TO TERMINALS IN AN APPROVED MANNER, 4) BE PROPERLY SUPPORTED, AND 5) BE PROVIDED WITH SUITABLE SEALS WHERE NECESSARY. SEE NEC SECTION 502-12.

G. RECEPTACLES AND ATTACHMENT PLUGS USED IN CLASS II, DIVISION 1 AREAS ARE REQUIRED TO BE APPROVED FOR CLASS II LOCATIONS AND PROVIDED WITH A CONNECTION FOR AN EQUIPMENT GROUNDING CONDUCTOR.

IN DIVISION 2 AREAS, THE RECEPTACLE MUST BE DESIGNED SO THE CONNECTION TO THE SUPPLY CIRCUIT CANNOT BE MADE OR BROKEN WHILE THE PARTS ARE EXPOSED. THIS IS COMMONLY DONE WITH AN INTERLOCKING ARRANGEMENT BETWEEN A CIRCUIT BREAKER AND THE RECEPTACLE. THE PLUG CANNOT BE REMOVED UNTIL THE CIRCUIT BREAKER IS IN THE OFF POSITION, AND THE BREAKER CANNOT BE SWITCHED TO THE ON POSITION UNLESS THE PLUG IS INSERTED IN THE RECEPTACLE. SEE NEC SECTION 502-13.

NOTE: NEC-NATIONAL ELECTRICAL CODE, NFPA 70.

TABLE 7 (CONTINUED)

H. SIGNALING SYSTEMS AND OTHER SIMILAR SYSTEMS: SEE NEC SECTION 502-14.

I. EQUIPMENT GROUNDING IS REQUIRED OF ALL NON-CURRENT-CARRYING METAL PARTS OF THE ELECTRICAL SYSTEM. LOCK NUTS AND BUSHINGS MUST NOT BE RELIED UPON FOR ELECTRICAL CONNECTION BETWEEN RACEWAYS AND EQUIPMENT ENCLOSURES. IF LOCKNUTS OR BUSHINGS ARE USED, BONDING JUMPERS ARE REQUIRED. SEE NEC SECTION 502-16.

NOTE: NEC-NATIONAL ELECTRICAL CODE, NFPA 70.

IN GENERAL, EQUIPMENT IN CLASS II, DIVISION 1 LOCATIONS SHOULD BE DUST-IGNITION PROOF, WHILE EQUIPMENT IN DIVISION 2 LOCATIONS NEED ONLY BE DUST TIGHT. ADDITIONALLY, EQUIPMENT SHOULD BE ABLE TO FUNCTION AT FULL RATING WITHOUT CAUSING EXCESSIVE DEHYDRATION OR CARBONIZATION OF ORGANIC DUST DEPOSITS. MAXIMUM OPERATING SURFACE TEMPERATURES ARE GIVEN IN TABLE 8. SINCE SOME GROUP G CHEMICAL AND PLASTIC DUSTS HAVE IGNITION TEMPERATURES APPROACHING OR BELOW THOSE GIVEN IN THE TABLE, EQUIPMENT USED WITH SUCH DUSTS SHOULD HAVE EVEN LOWER OPERATING SURFACE TEMPERATURES.

TABLE 8. MAXIMUM SURFACE TEMPERATURES

	EQUIPMENT THAT IS NOT SUBJECT TO OVERLOADING		EQUIPMENT (SUCH AS MOTORS OR POWER TRANSFORMERS) THAT MAY BE OVERLOADED			
			NORMAL OPERATION		ABNORMAL OPERATION	
CLASS II GROUP	DEGREES °C	DEGREES °F	DEGREES °C	DEGREES °F	DEGREES °C	DEGREES °F
E	200	392	200	392	200	392
F	200	392	150	302	200	392
G	165	329	120	248	165	329

SOURCE: TABLE 502-1 OF THE NATIONAL ELECTRICAL CODE, NFPA 80-1981.

THE FOLLOWING SECTIONS FURTHER EXPLAIN THE REQUIREMENTS FOR TRANSFORMERS AND CAPACITORS, SEALING, PIPE VENTILATION, AND GROUNDING AND BONDING IN CLASS II HAZARDOUS LOCATIONS.

TRANSFORMERS AND CAPACITORS

IN CLASS II, DIVISION 1 LOCATIONS, ALL TRANSFORMERS AND CAPACITORS MUST BE INSTALLED IN VAULTS OR MUST BE APPROVED AS A COMPLETE ASSEMBLY FOR CLASS II LOCATIONS. IN DIVISION 2 AREAS, TRANSFORMERS AND CAPACITORS CONTAINING LIQUIDS THAT WILL BURN MUST BE INSTALLED IN A VAULT. HOWEVER, NO TRANSFORMER OR CAPACITOR MAY BE INSTALLED WHERE ALUMINUM, MAGNESIUM, OR OTHER METALS OF SIMILARLY HAZARDOUS CHARACTERISTICS MAY BE PRESENT.

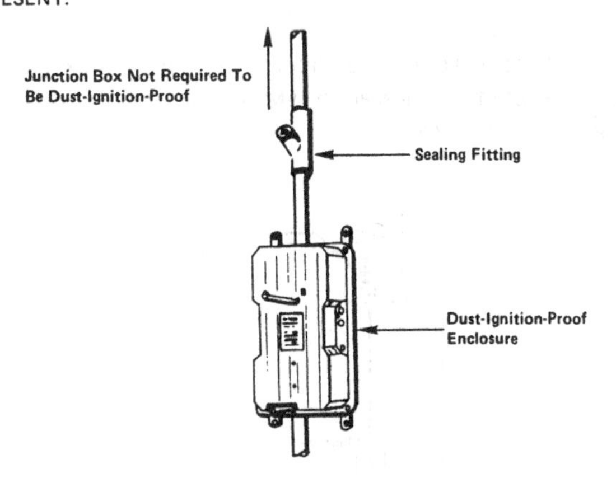

WHEN DUST-IGNITION-PROOF ENCLOSURES ARE IN A DUST-HAZARD AREA (CLASS II DIVISION 1 AND DIVISION 2) AND ARE CONNECTED BY A RACEWAY TO A NON-DUST-IGNITION-PROOF ENCLOSURE WHICH IS STILL IN A CLASS II LOCATION, DUST MUST NOT GET INTO THE APPROVED ENCLOSURE THROUGH THE RACEWAY. (NEC SECTION 502-5)

FIGURE 90. PREVENTING DUST FROM ENTERING THE DUST-IGNITION-PROOF ENCLOSURE BY SEALING BETWEEN ENCLOSURES

THIS CAN BE ACCOMPLISHED IN ONE OF THE FOLLOWING WAYS:

(1) BY INSTALLING PERMANENT, EFFECTIVE SEALS WITH FITTINGS THAT ARE EASY TO REACH FOR REPAIRS. SEE FIGURE 90.

(2) BY ARRANGING 10-FOOT OR LONGER RACEWAYS HORIZONTALLY BETWEEN ENCLOSURES. SEE FIGURE 91.

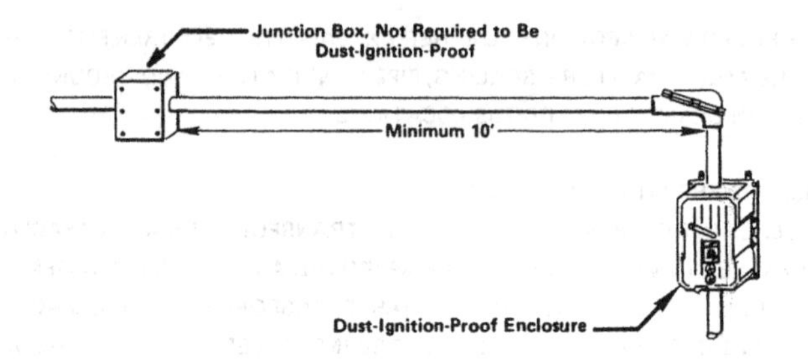

FIGURE 91. PREVENTING DUST FROM ENTERING THE DUST-IGNITION-PROOF ENCLOSURES BY HORIZONTAL DISTANCE (NO SEAL)

(3) ARRANGING 5-FOOT OR LONGER VERTICAL RACEWAYS THAT EXTEND DOWNWARD FROM THE DUST-IGNITION-PROOF ENCLOSURE TO A GENERAL PURPOSE ENCLOSURE. SEE FIGURE 92.

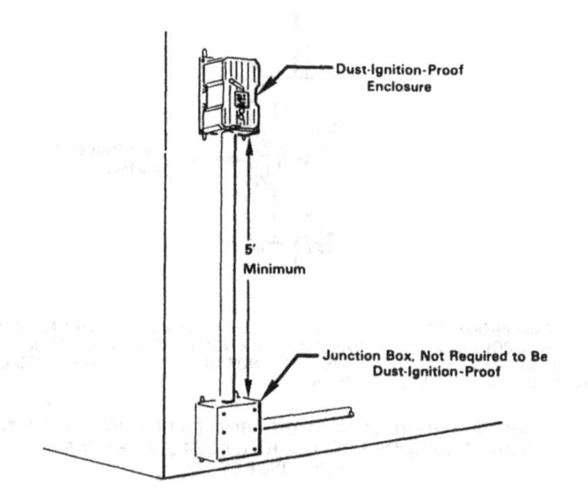

FIGURE 92. PREVENTING DUST FROM ENTERING THE DUST-IGNITION-PROOF ENCLOSURE BY VERTICAL DISTANCE (NO SEAL)

PIPE VENTILATION

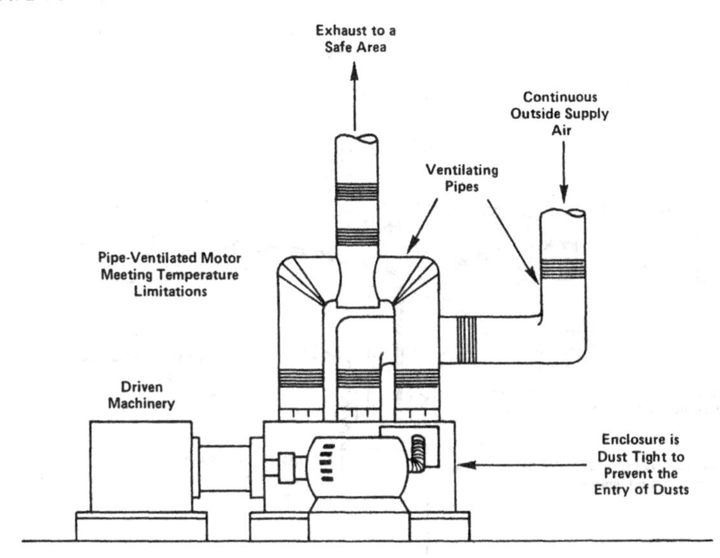

PIPE-VENTILATED MOTORS, GENERATORS OR OTHER ROTATING ELECTRICAL MACHINERY MUST BE ENCLOSED IN A DUST-TIGHT ENCLOSURE THAT IS CONNECTED TO OUTSIDE CLEAN AIR.

IN CLASS II DIVISION 1 LOCATIONS, VENTILATING PIPE MUST BE DUST-TIGHT – THAT IS, CONSTRUCTED TO MINIMIZE THE ENTRANCE OF DUST. SEE NEC SECTION 502-9(a).

IN CLASS II DIVISION 2 LOCATIONS, VENTILATING PIPE MUST BE TIGHT ENOUGH TO PREVENT THE ENTRANCE OF APPRECIABLE QUANTITIES OF DUST AND TO PREVENT SPARKS AND BURNING MATERIAL FROM ESCAPING. SEE NEC SECTION 502-9(b).

FIGURE 93 ILLUSTRATES A PIPE-VENTILATED MOTOR FOR CLASS II DIVISION 1 AND 2 AREAS.

FIGURE 93. TOTALLY ENCLOSED PIPE-VENTILATED MOTOR

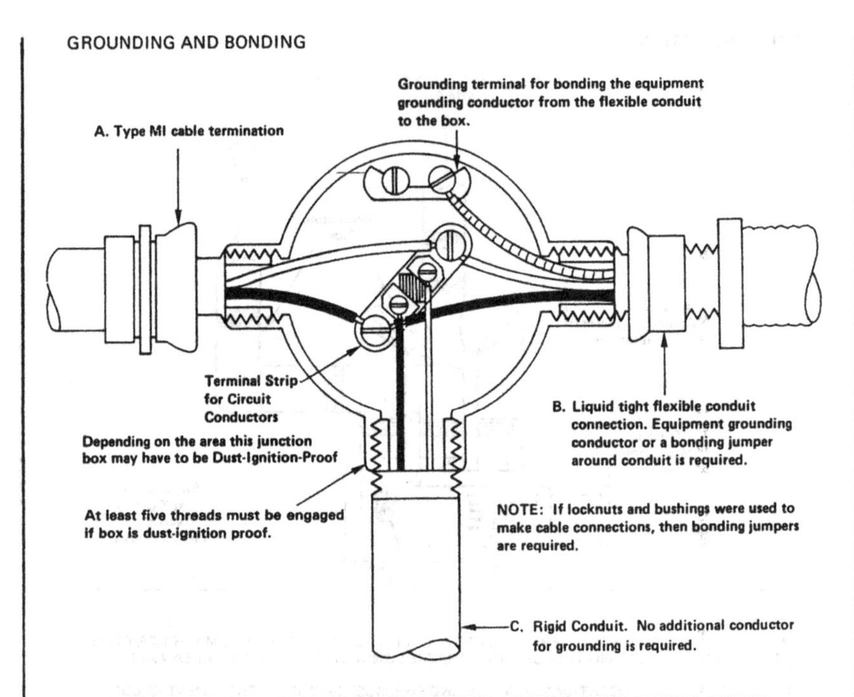

IN CLASS II LOCATIONS ALL EXPOSED NON-CURRENT-CARRYING METAL PARTS OF THE ELECTRICAL SYSTEM MUST BE GROUNDED. BONDING JUMPERS ARE USED TO PREVENT ARCS ACROSS JOINTS AND ASSURE GROUNDING AROUND FLEXIBLE CONNECTIONS.

FIGURE 94. BONDING IN CLASS II HAZARDOUS (CLASSIFIED) LOCATIONS

CLASS III HAZARDOUS LOCATIONS

CLASS III HAZARDOUS LOCATIONS ARE AREAS WHERE IGNITABLE FIBERS AND FLYINGS ARE PRESENT. IN GENERAL, EQUIPMENT ACCEPTABLE FOR USE IN CLASS II, DIVISION 2 LOCATIONS IS ALSO ACCEPTABLE FOR INSTALLATION IN CLASS III LOCATIONS. EQUIPMENT IN CLASS III LOCATIONS SHOULD BE ABLE TO OPERATE AT FULL RATING WITHOUT CAUSING EXCESSIVE DEHYDRATION OR CARBONIZATION OF ACCUMULATED FIBERS OR FLYINGS. THE MAXIMUM OPERATING SURFACE TEMPERATURE IS 165° C (329° F) FOR EQUIPMENT THAT IS NOT SUBJECT TO OVERLOADING, AND 120° C (248° F) FOR EQUIPMENT THAT MAY BE OVERLOADED.

FIGURE 95 AND TABLE 9 SUMMARIZE SOME OF THE REQUIREMENTS FOR INSTALLATIONS IN CLASS III LOCATIONS.

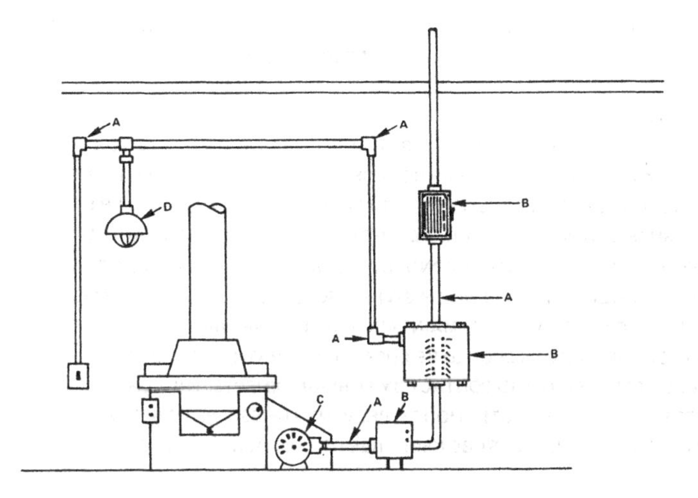

FIGURE 95. CLASS III HAZARDOUS LOCATIONS
(SEE TABLE 9)

TABLE 9. SUMMARY OF CLASS III HAZARDOUS LOCATIONS

(SEE FIGURE 95)

A. IN CLASS III HAZARDOUS LOCATIONS, WIRING MUST BE WITHIN A THREADED METAL CONDUIT OR BE OF TYPE MI OR MC CABLE UNLESS FLEXIBILITY IS REQUIRED. FITTINGS AND BOXES ARE REQUIRED TO PROVIDE AN ENCLOSURE WHICH WILL PREVENT THE ESCAPE OF SPARKS OR BURNING MATERIAL. SEE NEC SECTION 503.3

B. SWITCHES, CIRCUIT BREAKERS, MOTOR CONTROLLERS, AND SIMILAR DEVICES USED IN CLASS III HAZARDOUS LOCATIONS MUST BE WITHIN TIGHT METAL ENCLOSURES THAT ARE DESIGNED TO MINIMIZE THE ENTRY OF FIBERS AND FLYINGS AND MUST NOT HAVE ANY OPENINGS THROUGH WHICH SPARKS OR BURNING MATERIALS MIGHT ESCAPE. SEE NEC SECTION 503-4

C. MOTORS, GENERATORS, AND OTHER ROTATING ELECTRIC MACHINERY MUST BE TOTALLY ENCLOSED NONVENTILATED, TOTALLY ENCLOSED PIPE-VENTILATED, OR TOTALLY ENCLOSED FANCOOLED. THE WINDINGS OF TOTALLY ENCLOSED NONVENTILATED MOTORS ARE COMPLETELY ENCLOSED IN A TIGHT CASING AND ARE COOLED BY RADIATION AND CONDUCTION THROUGH THE FRAME. ENCLOSED PIPE-VENTILATED MOTORS HAVE OPENINGS FOR A VENTILATING PIPE, WHICH CONVEYS AIR TO THE MOTOR AND THEN DISCHARGES THE AIR TO A SAFE AREA. SEE FIGURE 93. IN TOTALLY ENCLOSED FAN-COOLED MOTORS, THE WINDINGS ARE COOLED BY AN INTERNAL FAN THAT CIRCULATES AIR INSIDE THE ENCLOSURE. UNDER CERTAIN CONDITIONS, SELF-CLEANING TEXTILE MOTORS AND STANDARD OPEN-TYPE MACHINES MAY BE USED. (SEE NEC SECTION 503-6.)

D. LIGHTING FIXTURES MUST HAVE ENCLOSURES DESIGNED TO MINIMIZE THE ENTRY OF FIBERS, TO PREVENT THE ESCAPE OF SPARKS OR HOT METAL, AND TO HAVE A MAXIMUM EXPOSED SURFACE TEMPERATURE OF LESS THAN 165° C. (NEC SECTION 503-9)

NOTE: NEC-NATIONAL ELECTRICAL CODE, NFPA 70.

IN ADDITION, REQUIREMENTS PERTAINING TO CRANES AND HOISTS, EXPOSED LIVE PARTS, AND GROUNDING ARE SUMMARIZED AS FOLLOWS:

CRANES AND HOISTS

ELECTRIC CRANES, HOISTS, AND SIMILAR EQUIPMENT INSTALLED OR LOCATED TO OPERATE OVER AREAS WHERE COMBUSTIBLE FIBERS ARE PRESENT MUST HAVE AN UNGROUNDED POWER SUPPLY THAT IS ISOLATED FROM ANY OTHER SYSTEM. ALSO, SUCH EQUIPMENT MUST HAVE A MEANS OF ALARMING AND AUTOMATICALLY DE-ENERGIZING THE CONTACT CONDUCTORS WHEN A GROUND FAULT OCCURS. A GROUND FAULT INDICATOR WHICH GIVES VISUAL AND AUDIBLE ALARM IS ALSO ACCEPTABLE IF THE ALARM IS MAINTAINED UNTIL THE CIRCUIT IS OPENED. THE CONTACT CONDUCTORS SHOULD BE LOCATED SO THAT THEY ARE GUARDED AGAINST TAMPERING AND CONTACT BY FOREIGN OBJECTS. THE CURRENT COLLECTORS MUST HAVE PROTECTION TO PREVENT THE ESCAPE OF SPARKS OR HOT PARTICLES, AND THEY MUST BE KEPT FREE OF LINT ACCUMULATIONS (SEE NEC SECTION 503-13.)

LIVE PARTS

LIVE PARTS OTHER THAN CONTACTS AND COLLECTORS FOR CRANES AND HOISTS MAY NOT BE EXPOSED IN CLASS III LOCATIONS.

GROUNDING

GROUNDING REQUIREMENTS FOR CLASS III LOCATIONS ARE THE SAME AS THOSE FOR CLASS II LOCATIONS (SEE NEC SECTIONS 503-16 AND 502-16).

(c) Conduits. All conduits shall be threaded and shall be made wrenchtight. Where it is impractical to make a threaded joint tight, a bonding jumper shall be utilized.
(d) Equipment in Division 2 locations. Equipment that has been approved for a Division 1 location may be installed in a Division 2 location of the same class and group. General-purpose equipment or equipment in general-purpose enclosures may be installed in Division 2 locations if the equipment does not constitute a source of ignition under normal operating conditions.

Index

9781468466225